Product Training for the Technical Expert

Product Training for the Technical Expert

The Art of Developing and Delivering Hands-On Learning

Daniel W. Bixby

Registered Office(s)
John Wiley & Sons, Inc., 111 River Street, Hoboken, NJ 07030, USA

John Wiley & Sons Ltd, The Atrium, Southern Gate, Chichester, West Sussex, PO19 8SQ, UK

Editorial Office
The Atrium, Southern Gate, Chichester, West Sussex, PO19 8SQ, UK

For details of our global editorial offices, customer services, and more information about Wiley products visit us at www.wiley.com.

Wiley also publishes its books in a variety of electronic formats and by print-on-demand. Some content that appears in standard print versions of this book may not be available in other formats.

Library of Congress Cataloging-in-Publication Data

Names: Bixby, Daniel W., 1972– author.
Title: Product training for the technical expert : the art of developing and delivering
   hands-on learning / by Daniel W. Bixby.
Description: First edition. | Hoboken, NJ, USA : John Wiley & Sons, Inc., [2018] | Series: IEEE |
   Includes bibliographical references and index. | Identifiers: LCCN 2017042183 (print) |
   LCCN 2017057183 (ebook) | ISBN 9781119260370 (pdf) | ISBN 9781119260387 (epub) |
   ISBN 9781119260349 (pbk.)
Subjects: LCSH: Manual training. | Product demonstrations. | Occupational training.
Classification: LCC TT165 (ebook) | LCC TT165 .B59 2018 (print) | DDC 658.3/12404–dc23
LC record available at https://lccn.loc.gov/2017042183

Cover design by Wiley
Cover image: © Laborant/Shutterstock

Set in 10/12pt Warnock by SPi Global, Pondicherry, India

Dad,
More than anyone I know, you believed that
doing *teaches better than* telling.
I love you.

Praise for *Product Training for the Technical Expert*

Most organizations have some of their greatest competitive advantage and wisdom locked hopelessly in the mind of their technical experts. Sure, as a force-of-one, subject matter experts get the job done incredibly well. But try to extend and multiply that individual expertise through learning and development efforts for others and the results are usually disappointing. Dan Bixby has been there and through this comprehensive and practical book, he shows how to overcome this challenge and unlock individual smarts for highly effective learning and development efforts. Further, as we are in the midst of a generational shift in the workforce, this is a very timely guide to ensure "what we know best" is effectively passed along to the new generation of talent.

Kevin D. Wilde
Executive Leadership Fellow,
University of Minnesota Carlson School of Management,
Former Chief Learning Officer, General Mills
Author of Dancing with the Talent Stars: 25 Moves That Matter Now

I was pleased to review Dan's new book—pleased because he addresses an old topic in a new way. He is making no assumptions for trainers who are not fully experienced and seasoned. He takes them step-by-step through practical and realistic methods to set up training graduates to actually be on-the-job performers. Enjoy, learn and be inspired.

Jim Kirkpatrick, PhD
Senior Consultant, Kirkpatrick Partners
Co-author of Kirkpatrick's Four Levels of Training Evaluation

Dan Bixby's book will help technical experts create training that actually works to improve performance. Backed by years of experience, Bixby connects with practical advice and empathy—and helps experts avoid the most common mistakes in product training.

Will Thalheimer, PhD
President, Work-Learning Research, Inc.
Author of Performance-Focused Smile Sheets:
A Radical Rethinking of a Dangerous Art From

Great book! A practical, must-have resource for any technical expert tasked with presenting information or training others. Bixby breaks the process down into simple, easy-to-follow steps that can be utilized by anyone to create dynamic and engaging learning sessions.

G. Riley Mills
co-author of The Bullseye Principle

Daniel Bixby's approach to product training for subject matter experts is practical, relevant, and exactly what anyone who is required to train others on technical content needs. He writes with candor and with a sense of ease, making the reader feel as though he is right there with you helping to develop your training competency. A must read for anyone on your team required to provide technical training to others!

Jennifer Alfaro
Chief Human Resources Officer

Recognizing that changing other people's behavior is the hardest thing you will ever try to do in your organization is the first step. The next step is recognizing that knowing the what, why, and how; telling the what, why, and how; and showing the what, why, and how is very different from facilitating learning. Dan has captured all of the key factors that one should consider in the conversion process from expert to learning facilitator. The ability to frame up the expertise into memorable, applicable, and measurable performance objectives is a complex, complicated endeavor. Dan's innovative and iterative approach can serve as a blueprint to do this successfully.

Kristin Ford Hinrichs
Chief Effectiveness Officer, Best In Learning

Product Training for the Technical Expert is a must read for anyone who endeavors to bring their training capabilities to the next level. Especially helpful to me was understanding the clear difference between a Presentation and Training. All too often we confuse the two, and frustrate both the learner and the facilitator.

John Schmidt
VP Fiber Market Development, CommScope, Inc.

This book is a must-read for anyone involved in teaching and training regardless of industry. The author, Dan Bixby, outlines principles that will transform any training program into a hands-on learning experience. I highly recommend this book for anyone who desires to improve learning and engagement within their organizations.

James M. Hunter, PhD
Chief Academic Officer, Emerge Education, LLC.

I have already begun recommending this practical manual to experts, trainers, training managers, and instructional designer who are interested in helping people acquire technical skills that are transferable to the workplace. It is obvious that Dan Bixby has enormous expertise and experience in the technical training field. Surprisingly (and fortunately), he has not forgotten the empathy required to engage the beginner's mind.

Sivasailam (Thiagi) Thiagarajan
Resident Mad Scientist, The Thiagi Group.
Author of Interactive Techniques for Instructor-Led Training.

Contents

Foreword

How do you design, develop, and deliver effective product training? Effective product training is more important than ever in government, business-to-consumer, and business-to-business environments.

Many technical professionals are asked during their career to train others on what they know. The request to train others is a compliment—people in your organization value product expertise and believe that those who best know a product are also those who can best train others.

There is a significant divide between knowing and doing. As you will learn in this book, the objective of product training is not knowledge transfer but rather proficiency development, as measured by the student's ability to *do*. Especially for adult learners, the goal is to enable the student to be comfortable using the product and confident specifying the product for others in their organization or for their customers.

So how do you design, develop, and deliver effective product training? This question is relevant to a range of professionals. You may be a field product expert being asked to train customers for the first time. You may be an individual in a call center asked to provide one-to-one or one-to-many on-demand product training through webinars. You may be responsible for corporate training and development organization— "training" is in your job title—and you are searching for more effective strategies to create training that results in improved measurable business outcomes. You may be a business leader focused on helping your customer—the technical product specifier— who is thirsty for product knowledge.

My job is a combination of all of the aforementioned roles, and my success directly depends on my organization's ability to develop and deliver effective product training. I lead the sales engineering organization for a multibillion dollar global leader in infrastructure solutions for communications networks with individuals in over 30 countries worldwide.

I have had the privilege to work with Dan Bixby and directly benefit from Dan's passion for product training excellence. Dan has developed a remarkable approach to developing product training and technical experts' skills to deliver product training. I have seen Dan's methodology work firsthand across cultures, languages, and geographies with my global sales engineering team.

You can benefit from the same methodology and principles detailed in this book. As you read this book, you will be surprised by many "ah-hah" moments where striking foundational ideas are conveyed. These insightful gems are critical in taking your product training effectiveness to new heights. A few examples you will encounter in this book are:

- The true objective of product training (product knowledge is important but is not the goal)
- Recognizing the difference between product training and product certification
- The journey to unconscious competence and how product experts should train audiences at different levels of expertise
- It's never *just* product training—always keep the business objectives in mind
- The 4 × 8 proficiency design model—content is important but must be developed *last*

Additionally, management professionals will benefit from "Executive Summaries" that recapitulate the learning objectives for book sections and highlight how you as a leader can support the individuals in your organization to learn and apply the methodology in this book.

As you, or those in your organization, put the principles in this book into practice, your ability to design, develop, and deliver effective product training will grow by leaps and bounds. Your customers will be confident about choosing and using your products, and you as a technical professional will now have skills, methodology, and appreciation to deliver effective product training.... You may even pick up Dan Bixby's passion for training excellence in the process.

Kevin Ressler, PhD
VP, Field Applications Engineering & Technical Centers

Preface

Nothing influences a human being to learn like another human being. Instructor-led training is not an old-fashioned approach only used by technology latecomers. As wonderful as the Internet is and as powerful as computers are, even the most advanced companies turn to their experts when it matters most. What they want is for the specialist to provide something beyond mere content. The human influence and perspective that a skilled practitioner provides can be more powerful than any interactive software or fancy delivery mechanism. It can also be dangerous. If you are that expert, you must deliver training that changes what people do with your product. Changing what they know about it is not enough.

Like me, you probably did not begin your career with the goal of becoming an instructor. My technical vocation began on the night shift, cleaning silicon wafers in a wet chemical lab. At the time, I thought my new profession was the end of a previous one—teaching. I was wrong. As my technology skills increased, so did my love for learning and helping others to learn. I have been blessed to be able to combine these two skills into a lifelong passion of educating adults in skills that will make them more successful in their careers.

My journey has led me to several companies where I have experienced everything, from being a one-man training department to leading global training programs for large corporations. Like you, I understand the unspoken pressure to be a technical expert first and a good instructor second. I want to help.

It has been my privilege throughout my career to transform engineers into artists. I have helped many people much smarter than I to become effective instructors by demonstrating to them that their students cannot be proficient in something they haven't done. I hope you will be next.

Acknowledgments

I have often skimmed this section and wondered why authors put down a list of names that many won't appreciate. Now I understand. Just as learning requires relationships, so does writing a book. These are just a few of those who have made this book possible.

My kids have been gracious to give up many of my evenings and weekends. Thank you. My brother, Bob, is probably glad this book is finally finished. Now we can spend our Tuesday calls talking about something else. Thank you for your patient ear. Mom, thank you for teaching me to read and write. I am grateful for both you and dad and the time you've spent assisting on this project. I learned so much about teaching from both of you. I'm glad Dad got to read the majority of this book, though I'm sad he never saw the finished product. What I wrote in the front of the book was what I had intended to write in his personal copy. Since he isn't here to read it, everyone gets to. To the rest of my family, both immediate and extended—many of whom became proofreaders by proxy—thank you for volunteering your skills.

Morten Tolstrup, thank you for pushing me to write and introducing me to Wiley. As an author yourself, you knew how much work this was going to be. Thank you for not telling me!

Dr Kevin Ressler, your feedback and encouragement was helpful and is appreciated. Jack Culbertson, I am so thankful for your help in making this book as practical as possible. It is better because of your input. Dianne Schanhaar, you have regularly encouraged me when I needed it most. Thank you.

Thank you, Rebecca Rosemeier and Sara Briggs for your help with the artwork used in this book.

So many others have encouraged me along the way. You have read and offered suggestions and feedback, corrected my grammar, or simply given a kind word of encouragement. I hesitate to mention some, knowing I will forget many: Thompson Lewis, Kristin Hinrichs, Jim Breezley, Christa Harrison, Jennifer Alfaro, Dave Gilbert, Dr David Farrar, Genevieve Farrar, Dr Will Thalheimer, Dr Brian Hanson, and Dr Jim Kirkpatrick. Thank you.

Finally, I'm not even sure how to write a proper thank-you to my wife, Brenda. I don't think I will. It would require too many pages. I'll find a better way. I love you.

How to Use This Book

And Who Should Read It

Self-Study

Instructors are learners. Leaders are learners. All successful employees are learners—they want to get better at the skills they have been asked to do, even if they seldom practice them. The goal of this book is to increase the effectiveness of anyone asked to provide hands-on learning.

If you are a product instructor, read the whole book. Take the time to answer the questions at the end of each chapter. You will find them helpful. There are questions about what you have just read and questions about what you are going to read. You may wonder why there are questions about a topic you have yet to read about. Two reasons. First, this encourages (as much as possible in a book) a two-way conversation. I don't know everything. Together, you and I can learn a lot more than we can alone. The answers you provide *before* you read the text will come from a very different perspective than the ones you provide *after* you read the text. This will give you a chance to compare them and learn more. Second, it is a teaching tool. By prompting you to think about one or two things first, I am encouraging you to look for those principles as you read. Feel free to use that technique with your own students!

Since product training is a team effort, there are many people who can benefit from parts of this book. Here are some examples, along with the sections they should concentrate on (Table i.1).

Group Study

Some technical training groups may choose to use this book as a workbook for group study. Take a chapter a week (or however often you meet) and come together ready to discuss (and maybe even debate!) the questions at the end of the chapter. Create your own questions as well and discuss how you can apply or have applied the principles in real life.

Textbook

I teach a class based on the principles in this book—a train-the-trainer course for product specialists. In general, I concentrate on certain elements in Parts one, three and four. If you find this book helpful, it may work for you as well. If you choose to do so, please

Table i.1 Who should read this book.

	Part One: Foundation	Part Two: Strategy	Part Three: Structure	Part Four: Facilitation	Part Five: Operation	Executive summaries
Technical trainers	✓	✓	✓	✓	✓	
Product specialists	✓	✓	✓			
Instructional designers	✓	✓	✓		✓	
Training coordinators	✓				✓	
Training managers	✓	✓	✓	✓	✓	✓
Product managers	✓	✓				✓
Field engineers	✓		✓			
Human resources	✓				✓	✓
Executive leadership						✓

do me the favor of letting me know. I would love to hear how this book is helping other product trainers.

Encouragement

I love training. Maybe you do, too. Maybe you don't—you do it because you have to. Perhaps you are an excellent and engaging instructor. Perhaps you've been told you are boring. Wherever you are at, when it comes to product training, this book should encourage you.

To be effective, instructors don't have to know all there is to know about training any more than an engineer needs to know everything about a particular product or technology. But we can all improve.

Practical Performance Guide

This book is not an encyclopedia of everything there is to know about product training. It is a practical resource from one person's perspective. Yes, I have a lot of experience in training all over the world. But I don't know everything. In fact, the more I learn, the more I realize how little I know. I hope that this book can provide a real-world difference in the way you teach.

More than anything, I hope you will do more than just read this book. The theme of this book is that your students learn best by doing. In the same way, readers of *Product Training for the Technical Expert* will learn best by doing. If you don't get anything else from this book, remember this one thing:

> ***You can't be proficient in something you haven't done.***

About the Companion Website

Don't forget to visit the companion website for this book:

www.wiley.com/go/Bixby

There you will find valuable material designed to enhance your learning, including:

1) Exercises
2) Checklist
3) Action verbs
4) Certification verification
5) Making it practical
6) Revision numbering

Scan this QR code to visit the companion website

Introduction

When subject matter experts are assigned the task of delivering a product training class, it is not typically because they are great trainers or because they can hold an audience spellbound with their charismatic personality. Usually, it's because they know a lot or have access to a lot of content. Your familiarity with your product solutions gives you a distinct advantage as a technical trainer. But training is like a two-edged sword. You need more than product knowledge. You need to know how to facilitate learning for others. Once you are armed with both your product skills and teaching skills, you will be ready to transform good training into great training.

You may already be a natural instructor. If so, I hope that the principles and suggestions in this book will sharpen that natural ability, strengthen your resolve to continuously improve, and broaden your skillset with practical tips and tricks.

Maybe you are afraid of teaching or think you don't have the skillset. This book will help you become a better instructor as well. As an expert, you will need, at some point in your career, to help a subject matter beginner improve his or her skills. If you aren't asked to get in front of a classroom, you may be asked to assist a trainer, to give a webinar, to provide content for an eLearning module, or to mentor a new employee. The concepts in this book will help with all of that.

Of course, my hope is that seeing people become proficient in something because of your training will make you proud. I hope that it will make you and your company more successful and that training will become more than just something you do, but something you do better.

Being skilled at something implies that one has more than just knowledge. An accomplished artist combines knowledge with art to create something unique and exquisite. A skilled trainer does the same. You should know your product well. You must take that knowledge and create something unique and exquisite: an individual with a new skill.

It takes skill to teach skill.

Part I

The Foundation of Hands-On Learning

1

Hands-On Learning in the Classroom

Articulate Your Approach

You are a product expert.

I've worked with you. I've been on conference calls with you. I've heard those who respect your expertise in your product and say that you should be the one to train the beginner, because you know the product as well as anyone. I've seen you reluctantly agree to teach them again. Maybe they'll get it this time.

Okay, so I haven't worked with you, but I've worked with many like you, and here's what else I know about you. I know that you are very smart. People say that you "just get it" when it comes to your product, technology, or industry. You can probably install or use your product with your eyes closed. When people need help understanding the limitations of your product or stretching the possibilities of what it can do, you are one of the few that gets the call. What others think is difficult, you find to be simple. You like to fix it, install it, program it, commission it—whatever you do with your product—you like to do it yourself.

Now you've been told you need to "transfer your knowledge" to others.

But you're frustrated that no matter how much you go over it with people, they still think it's difficult to understand. It seems so obvious to you. You're afraid some might think you don't want others to become as smart as you are—that you're one of "those" knowledge-hoarding engineers—but that's not true. You genuinely want to make it easy for them as well.

"But," you say, "I can't."

Take heart. Here's the good news and the bad news all at once: you're right. You can't.

But what you can do is help them learn it for themselves. And that is a skill that starts right here.

Product Training as You Know It

I've been involved with product training for nearly 20 years, so I'm pretty confident that if you have taken many technical training courses delivered by subject matter experts, you are familiar with the training process described in the following text. Like the fictitious John in this scenario, you, too, may have struggled with the following progression.

Product Training for the Technical Expert: The Art of Developing and Delivering Hands-On Learning, First Edition. Daniel W. Bixby.
© 2018 John Wiley & Sons Ltd. Published 2018 by John Wiley & Sons Ltd.
Companion website: www.wiley.com/go/Bixby

John will deliver the content
John believes this is the most critical step. If he can only accomplish one thing, he must make sure his students get "the training," by which he refers to the information in his presentation. He hates to be rushed. He would love to explain it better, but he has only a limited time to deliver this content.

John will more thoroughly explain the content
This is what John calls a *real* training class, as opposed to a presentation (though, admittedly, he uses the same material for both). He has put as much information on the slides as he can, just in case he doesn't have enough time. That way his students can read them later. With a little more time, John can add stories and anecdotes about how he has used the product successfully, as well as a question and answer time at the end.

John will demonstrate the product
John loves this part, because he loves working with his product. He gets to show them how easy it is to use once you become proficient using it. He likes watching students' eyes light up in apparent understanding as he pushes buttons and adjusts settings at the front of the room. Since John wants to confirm their learning, he asks the class to gather around him, so they can observe what he is doing more closely.

John will let the students experience the product
John's classes don't often get this far, though he wishes they would. He believes in hands-on training—that the students should test their newfound knowledge on the equipment they've been studying. They should be able to change those settings and troubleshoot those problems, but the lab time will help to reinforce it. Not all of the students will get through the exercises, but if they struggle, John will be there to show them how to do it.

Have you been there? My guess is you have. Of course, you will need to adjust some things to your setting, or to an online or webinar environment, but the general progression is there. Technical experts, like John, have become so accustomed to the "Tell, Explain, Show, Try" model that they actually expect this progression when they go to a product training class (Figure 1.1). This undefined and ineffective template has become the accepted standard for product training. This model needs to change. It is hurting your learners. You didn't become an expert by merely receiving information. Your students won't either. You became an expert by doing it.

What Makes Training Effective?

Your training must be effective. In the words of a colleague and expert in training design and facilitation, Terri Cheney, "Ineffective training is expensive."[1] If you provide training that doesn't work, you are wasting your time and your students' time, and you are wasting their money and your company's money. No one wins unless the training works.

Figure 1.1 Ineffective product training process.

Most product training begins with a problem. The problem may be real, and it may be that training is part of the solution. I'm going to start with the assumption that other potential factors have been eliminated—the quality of the product, the environment, and so on are not issues, and the right product is being used for the right solution. The real need truly has been narrowed down to the way users understand the product functionality. Even then, simply providing a training class is never *the* solution. That may surprise you, coming from a training professional in a book about training.

The reality is that you already know that. You're a subject matter expert. It is precisely because you are a subject matter expert that you have been asked to provide the training or help develop the curriculum. You know that asking a subject matter beginner to give the training could be catastrophic. You agree with the statement that bad training on your product is worse than no training. "So," you're thinking "it's *good* training. When people need to understand something new, *good* training is the solution."

You're getting closer. The problem is that most subject matter experts and even most corporate executives are unable to define what good training is. Their emphases are placed on any number of good and very important things that actually detract from the goal. Following are a few that are prevalent in product training:

Content. By far the front-runner on this list is content. Bad content equals bad training. Therefore, we reason, good content must mean good training.

Unfortunately, this deduction is not true. Content is extremely important. It is true that bad content would turn into bad training and that having good content is essential. However, just because something is important—or even essential—doesn't make it *the* solution. Good content, by itself, is not the solution to your training needs. Understanding this principle is so important that it is the focal point of Chapter 8.

Presentation skills. No one wants to sit in a class or on a webinar with an instructor who can't communicate well. If instructors can take good content and add in good presentation skills, maybe *that* is the solution for good training.

Here again, good presentation skills are necessary. I love teaching subject matter experts how to present, but it is not *the* solution to your training needs.

Availability. Here, the emphasis is quantity over quality. The more knowledge one distributes to those using their products, the fewer issues they will have. Product experts need to spread their knowledge. If those experts can just take advantage of available technologies and get more people through the training course, that will solve any education problems.

Believe it or not, this is a common approach. The desire to make your training available to the most people possible is a great desire. Sometimes executives mandate a certain availability; other times it is the subject matter experts that believe availability is the answer. However, if you have determined to deliver your course via eLearning or webinar before you even develop the objectives for the course, you may be guilty of focusing on the wrong thing first.

The list could go on to include other great things, like adult education and profitability. Both of those, to some degree or another, are key pillars in a philosophy of proficiency, but none of them stand alone.

So, what is the solution to the problem? The solution is to effectively change a person's skills or behavior—to change how they *do* something. Everything else just helps you reach that goal. You need good content, great presentation skills, and a good mechanism to deliver training that will change behavior. You certainly want to improve the skillsets of as many people as possible. Making the outcome something that can be observed and measured, however, will help you choose the right content and deliver it the right way.

Your Goal: Proficiency

To be proficient in a skill or task, according to Merriam-Webster's simple definition, is to be good at doing something.[2] Fuller definitions might include words like expert, well advanced, or competent. The desire of all product trainers should be to help others become experts with their product—to be able to use their products with competence.

It is important to point out the verb "to do" in the definition of proficient. To be proficient in something requires more than just knowledge. It requires being able **to do something** with that knowledge. Knowing a lot about your product is useless. Doing something with your product can launch immense success.

Product proficiency training involves more than just teaching people about your product. It is about making the training effective. The training is only effective when it changes the way the learner uses your product. This is not an issue of using the right terminology. This is a vital foundation of effective training.

If you merely teach students *about* your product, you may well gain some admirers, but you will not produce specialists or skilled users. Only changing how people use or what they do with your product will produce skilled users. Skilled users are the lifeblood of a product. The more there are, the more successful the product will be. Knowledge may affect behavior, but when push comes to shove, behavior is always more important than knowledge.

If you are lying on the operating table, what is more important to you: a doctor who knows a lot about the anatomy or one who can effectively perform the surgery you need? What about the pilot flying your airplane? Which do you want—knowledge or proficiency? The same can be said of you and your products. What makes your product valuable is what it does. What makes you valuable is not how much you know about your product, but what you are able to do with your product. What will make you effective as an instructor is how successful you are at changing what others do with your product.

Articulating Your Training Approach

In many ways, defining your approach to training is like articulating a good vision or purpose statement. It takes time and effort to create one that is broad enough to cover any learning application, but specific enough to define the approach you should take and demand consistency. Your primary goal is to clearly state how you will teach your students. You need a blueprint that will guide you no matter the product on which you are training on and no matter who you are training.

But first, you must understand how your students learn. Think about your product. Imagine the difficulties you would have if it had been designed without any understanding of how it would be used after production. It is more likely that product managers and engineers have taken great lengths to understand the market, the environment, the user preferences, and other factors that affect how your customers will use your product. The foundation of your approach starts with an understanding of what you want to accomplish—in this case, teaching proficiency. It is possible for training programs without a documented focus on proficiency to be successful. They may teach proficiency without really trying. Chances are, though, that that sort of training will be hit or miss.

There are several practical benefits to having a defined philosophy. First, by defining what you expect the outcome to be, you will ensure consistent training. Second, by emphasizing proficiency over knowledge, you will increase retention. Proficiency endures longer than information, which can be easily forgotten. Third, and most importantly, you will guarantee that your training is effective.

Three Things to Document

You want your training to be effective. Every time. When you document three important areas of instructor-led training, it can be. The beauty of documenting how your students will achieve proficiency is that the process of documenting will provide answers to most of your questions before they are even asked.

Creating an effective product training class is like building a house. First, a house needs a well-designed strategy that includes blueprints and financial backing. Second, a house needs a strong and supportive structure that aligns with your plans. Third, the house must be livable and presentable or it won't get used. Consider how that applies to training:

1) Strategy. Training blueprints include the financial information that will make them successful as well as the expectation of what the finished product should provide.
2) Structure. Effective training doesn't just happen. Effective training is carefully built with a thoughtful and proven process.
3) Delivery. If training isn't presented in a way that encourages learning and is enjoyable for the learner, it won't get used. And it must get used since a great product that doesn't get used has no value.

Try a little exercise with me. You will be tempted to skip over it, but take the time to do the exercise, since it will help to shape your definition of effective training. Think about a class or learning module you have taken in the recent past—preferably one where you learned a lot and enjoyed the class. Write down answers to the following questions as you consider that learning experience.

Exercise 1.1 Class you have taken

Class you have taken
Why did the company or organization offer this class?

What can you do differently after this class?
How did you learn in this class?

I asked you to answer those questions before reading any further so that you can see later how a different approach may change some of the answers. Before taking a deeper look at the strategy, structure, and delivery of product training, take a moment to consider a class that you might teach or have recently taught on one of your products or a technology of your choice. Answer similar questions about the class, course, or module you are planning to teach.

Exercise 1.2   Class you will deliver

Class you will deliver
Why should your company or organization offer this class?
What should students who take this class be able to do differently after the class?
How will they learn in this class?

Refer back to these answers as you read through this book. Add or adjust your answers as necessary, but make sure you can answer them with confidence.

Adult Learning Principles: The Foundation of Hands-On Learning

Why, what, how? All three questions are really about how adults learn. Perhaps you thought the last question was a typo that I meant to ask *what* students would learn *about.* Maybe you thought it was a question about the delivery medium; will it be an eLearning class or will it be an instructor-led class? Both the topic and the delivery method are important. But the third question really is about *how*—how the information will be transformed into understanding—not about materials or delivery styles. Understanding how adults learn is an essential part of being a good instructor.

Training is a skill. There are good trainers and there are poor trainers. Facilitating technical learning is like the skill of photolithography inspection, web development, computer skills, engineering, or any other technical skill. It requires an understanding of the topic to get started and experience to become proficient. You don't expect a new technician to have the same expertise that you do, much less someone who has no technical background at all. Then why would you expect someone who hasn't studied how to teach or how adults learn to be a good instructor?

Training, like the technical skills I mentioned earlier, also requires another ingredient to make a great teacher. Behind every successful technical expert is a little bit of artistry. Part of what will make you a technical success is your ability to hear that the machine isn't working properly, to see a default before it is released, and to feel an imperfection. That's art, not science. Those are instincts that become part of who you are as a technician. However, those fine-tuned instincts also make it harder for you to teach the skill.

Adult education is a field that some spend their lives studying. You don't need that level of understanding, but you do need to understand some of the basics. You need to be able to answer the question, "How will students learn in this class?" The philosophy of product proficiency doesn't change. That said, the strategy and the structure of training might change as new reasons and ways to design and build effective training are developed. The delivery might change as well, as new technologies are made available to us. The principles of adult learning do not. As demonstrated in Figure 1.2, they are the foundation that holds the three pillars.

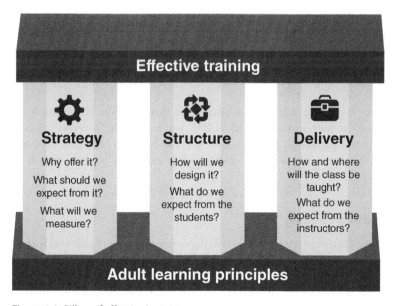

Figure 1.2 Pillars of effective learning.

The Strategy of Hands-On Learning

Why should your company or organization offer this class? The first question will drive you to define a purpose statement. Product solution training differs from general education in one key area: the impetus to have it in the first place. All product training is driven by a business strategy. Never offer a class unless you know why it provides value to your company and to your students. If you work in a for-profit business, your company's mission statement will likely be a major part of your purpose statement. Even if you work for an institution of higher learning or a nonprofit organization, it is critical to be able to understand and articulate the purpose statement.

Make certain your answer to the question earlier reflects your company's vision for your product. Perhaps you wrote down an answer like the one I often see: "So that students can learn more about our products." Or perhaps you were more specific in your answer: "So we don't get so many calls in technical support." That last one is a better answer, but it doesn't answer why it is important to reduce technical support calls. Your training should support the goal of the company. Usually, the main goal of the company is to sell more products. Sometimes, it is to save money. A better answer may be "to increase sales by $700,000" or "to decrease technical support spending by $200,000."

The Structure of Hands-On Learning

What should students who take this class be able to do differently after the class? The second question pertains to the expected outcome of your class. If you don't know what you are seeking to change, you have no framework to build on. Almost as importantly, it is impossible to be consistent if you cannot answer that question.

Product training that is consistently effective always completes the phrase, "At the end of this class, students will be able to do _____." Note that this phrase is specific, without ambiguous expressions like "know more" and "understand better." It is almost impossible to be consistent unless you are precise. When a former student cannot adjust a sensor on your product, his or her boss is not going to ask them if they learned "a little more" in your class. They are going to ask if the instructor teaches how to adjust that sensor in the class. "No" is an appropriate answer, but "sometimes" or "depending on the instructor" is not.

Often, your purpose statement will imply an answer to this question. For example, if your "why" answer was "to increase sales by $700,000," you might add "by ensuring our resellers can integrate our product." If your answer was "to decrease technical support spending by $200,000," you might add "by verifying that our installers can install it correctly."

The Delivery of Hands-On Learning

How and where will the classes be taught? How will the instructor ensure that real learning happens? Questions like these are important to develop a clean, organized teaching environment. A structurally sound and well-designed house will not be used if it is not livable and presentable. The same is true with training. A great training program, driven by a solid strategy and built on great theories, is still useless if those theories aren't actualized in the delivery of the learning. The best way for subject matter experts to do that is to abandon the idea of presenting training and adopt, instead, the concept of facilitating learning.

Facilitating effective learning is what this book is about, but getting there requires some groundwork. You can't hang pictures in your house before the walls are built. Too many instructors have been told it is their presentation skills that need to improve, when the structure of the training course is the real problem or when the lack of a forward-thinking strategy would doom the training anyway.

But it is important to hang those pictures—to perfect those delivery skills. Training that is boring will not attract students. More importantly, however, your training must be effective. It must help your company meet its goals by changing what people *do* with your product.

Conclusion

It is challenging enough to teach one effective and well-delivered class. Teaching effectively in every class or creating a program that involves multiple instructors all seeking to teach effective classes requires enormous effort. The key to your success will be determined by how you answer those questions. Are you confident that this course will help your company achieve its goals? Can you say, with reasonable certitude, that all students who successfully complete this course will be able to perform the intended task?

You will never be able to do everything. You will have decisions to make: how many students to include, who can take the class, what technologies will you use, how will you measure results—all are discussions you must have before you offer training. If you don't, you'll end up altering what you do with every class. Inconsistent programs are difficult to improve. You will naturally reflect on your best examples (when you did teach about that sensor) and fail to improve the others.

Having the right approach to product education is just the beginning, but it is the right place to start. When you consider your product training, can you articulate what makes it effective? Can you answer the questions that define your approach to proficiency learning? Your ability to meet those two objectives is an important first step in becoming an effective instructor of your product, solution, or technology.

Making It Practical

The principles of teaching for proficiency instead of merely increasing knowledge can change the way we think about product training. How does it apply in your company or department?

1) Define how proficiency with your product is different than knowledge of your product.

2) In your own words, describe whether you would prefer the pilot of your airplane to be knowledgeable or to be proficient and why.

3) What are three areas that are important to articulate in order to provide consistently effective product training?

   a)

   b)

   c)

Before you read Chapter 2, "Experiencing Learning: Emphasize Skill over Information," answer these two questions:

1) In your own words, what do you think I mean by the phrase, "Don't teach to yourself"?

2) Write down how you think you learn best in one to three sentences.

Notes

1 Terri, Cheney. From Know to Do: A Quick Guide for Experts Training Other Adults. Unpublished.
2 Merriam-Webster online dictionary. http://www.merriam-webster.com/dictionary/proficient (accessed August 8, 2017).

2

Experiencing Learning

Emphasize Skill over Information

Do you do things because you have learned how to do them, or have you learned how to do them because you did them?

This, and similar questions about how humans learn, has kept philosophers and professors busy for centuries. I appreciate their desire to understand the details of how we learn. The simple answer to the question, especially when applied to product learning, is probably "Yes." Think about something you do with relative ease, like riding a bike or heating up a cup of coffee in the microwave. You do it because you learned how to do it, but you also learned how to do it because you did it. It is the continuous tango between thinking and doing, the DNA-like weave between the possibility and the encountered, and the scissor blades of could-be and has-been that create learning in human beings. Both must exist for true learning—the change in one's knowledge, skills, values, or worldview—to take place. But one—information—tends to get emphasized in the classroom, while the other—experience—gets left to chance. If your training is going to be effective, that must change.

How Does One Develop a Skill?

I challenge you to think about a skill that you can do now without really even needing to think about. Here are a few examples.

Exercise 2.1   Example skills

- Using the remote control
- Brushing your teeth
- Setting your alarm clock
- Downloading a smartphone app
- Playing a sport
- Changing a car tire
- Painting a room

- Making breakfast
- Riding a bike
- Dancing
- Making a purchase online
- Setting the table
- Connecting to the printer
- Playing a musical instrument

- Ironing your shirt
- Mowing the lawn
- Filling up and paying for gasoline
- Using a calculator
- Fixing a plumbing leak
- Installing a 3-way switch
- Editing a digital photo

As you consider the skill, think about the first time you did it. Did you do it perfectly the first time? Did you know everything there is to know about it before you first did it? I'm quite sure you didn't! The Cambridge Dictionary defines skill as "an ability to do an

Product Training for the Technical Expert: The Art of Developing and Delivering Hands-On Learning, First Edition. Daniel W. Bixby.
© 2018 John Wiley & Sons Ltd. Published 2018 by John Wiley & Sons Ltd.
Companion website: www.wiley.com/go/Bixby

activity or job well, especially because you have practiced it."[1] In other words, if you—or anyone, for that matter—could do it perfectly the first time, it would not be a skill.

Niels Bohr, a Danish physicist, is said to have stated that "an expert is someone who has made all the mistakes which can be made, in a narrow field." Indeed, if you have not made mistakes in your field, you are likely not an expert. I'm sure that he made the comment somewhat tongue-in-cheek, or perhaps out of an attempt to display humility. Of course, he was not advocating that mistakes alone make an expert. Instead, the pithy statement drives home that all of us—expert or novice—must allow even the negative experiences to change the way we behave. Simply put, we must submit ourselves to a lengthy process of attempts and failures, small incremental improvements, study, feedback, and practice.

Remember How You Became an Expert

Teaching others how to be an expert is much easier if you consider how you became one. I could offer to analyze over a century of adult learning theories, or I could sum it up in one extremely profound statement. You may want to write this deep thought down.

You cannot be proficient in something you haven't done.

Stunning, I know. But understanding that fact will change the way you teach. Teaching will become less about how much knowledge you can transfer and more about how you can help students relate to the information so well that they construct true understanding for themselves.

Learning a skill has an added dimension—that of aptitude—that must be distinguished from acquiring knowledge. *Teaching* a skill also has an added dimension. You cannot claim to know how to ride a bike if you have never ridden one. You cannot claim to have taught someone how to ride a bike if they haven't ridden it yet.

Learning is the thread that sews knowledge and proficiency together. But to teach that effectively, you must get away from the notion that there is a mandated sequence in which your students can learn. You don't teach anyone—adult or child—how to ride a bike by giving them a lecture on wheels and spokes or balance and brands. Your students might ace that quiz, but if they've never gotten on a bicycle and ridden it, they know less about it than my 6-year-old neighbor who just flew by. You cannot send your students out of the classroom and claim that they are proficient if they have never actually used the product.

Build on Your Students' Experiences

Your students have experiences, too. Regardless of the product that you are teaching others about, the learners in your classroom will come with at least some prior learning. They will bring their previous experiences into your classroom. The experience doesn't even have to be one directly with your product. Chances are there are many parallel areas of study that will affect how quickly or how well your students learn.

As in any teaching environment, it is important to remember that you are teaching individuals. Your students are not robots that can be programmed *en* masse. Think about what it would be like to teach two people from very different backgrounds how to drive a car. The first individual grew up riding in a car. Her mother would explain certain things to her, even when she was a child. She would ask questions like "Why are you pushing that?" The answer would be at her level, "When I push that it makes a light blink so that people know I'm going to turn here." On and on the learning goes, until 1 day she is driving you! The second individual, however, has never seen a car. He doesn't know what it does, much less how to control it. You can imagine that the learning experience would be very different for this individual, as would the teaching experience for the instructor.

Create Experiences in the Classroom

If a student already knows something or has experienced it in the past, it is possible to remind them of that learning simply by lecturing or demonstrating what you want to teach them. However, if the topic is completely new to them, they will need to experience it, or draw from previous experiences or interactions with the topic in order to learn it.

All product specialists know this to be true. The great question is why so few apply it to the event of training. Most will apply this concept by asserting (the truth) that true training is more than just an event. It takes on-the-job experience or practice to really become proficient in something. Few will argue with that. But you can speed up the learning process by creating an initial learning event that is experiential by design. Many believe that cognitive learning is learned while sitting and listening to an instructor, while skills are only learned on the job. You can change that.

On-the-job training is essential. Practice is important. What I'm advocating here is combining the learning and the doing. I'm suggesting that experiential learning is the only real way to teach a skill, even in the classroom, over a webinar, or in an eLearning module. Later I'll explore some practical ways to make this happen, but before then, let's try a little experiment.

Take a pencil and a ruler and connect the "1" dots to each other in the graph in Exercise 2.2.

Exercise 2.2 Numbered graph

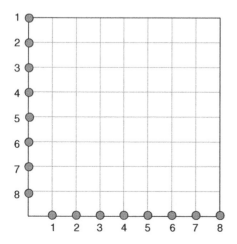

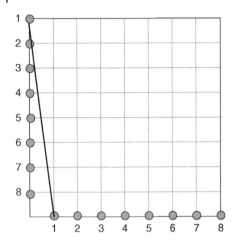

Figure 2.1 Graph illustration.

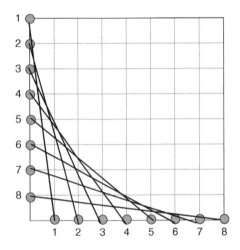

Figure 2.2 Parabolic curve.

Does it look like Figure 2.1?

Good! Now, go back to the first graph and connect the "2" dots in the graph (in Exercise 2.2). Then connect the "3" dots. Do the same for the 4, 5, 6, 7, and 8 dots.

Now, does it look like Figure 2.2?

Now, you may know what you just created from previous experience. But even if you don't, without any explanation, you have created a parabolic curve. You did so without a long explanation or demonstration; you did so on your own. Next time you train on your product, see if you can get the students to do what you are training them to do *while* you are teaching them. You'll discover that the excitement of learning will significantly increase. But even more importantly, you will discover that actual learning will increase as well.

Let Them Learn from Negative Experiences

But let's be honest. Sometimes your students will try to do something and it won't work. Not all experiences are equal. There are positive experiences and negative ones, and either of those can produce accurate learning or erroneous learning. As you can see from

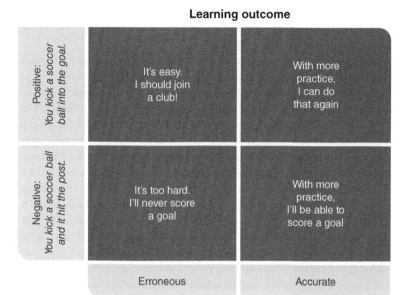

Learning outcome

Positive: *You kick a soccer ball into the goal.*	It's easy. I should join a club!	With more practice, I can do that again
Negative: *You kick a soccer ball and it hit the post.*	It's too hard. I'll never score a goal	With more practice, I'll be able to score a goal
	Erroneous	Accurate

Figure 2.3 Learning outcomes.

the grid in Figure 2.3, positive learning experiences can still generate erroneous learning outcomes, and negative learning experiences can generate accurate learning outcomes.

Allow Students to Make Mistakes

As an instructor, you hope to create as many positive learning experiences as possible, as long as they produce accurate outcomes. A controlled "negative learning experience" that produces an accurate learning outcome (I pushed this button and the power went out!) is still a positive experience since you designed the class to happen that way. However, unplanned negative experiences *can* happen in the classroom. Usually they are due to mistakes or faulty products. The challenge to the instructor is to turn them into accurate learning.

One of the roadblocks technical experts have in allowing novices to use their products is the fear that they will make mistakes. They will. But that's not always a bad thing. One way that our brain constructs knowledge is through the process of elimination. When your students are in a controlled environment, allowing them to do something wrong and learning from the mistake can be valuable. The best place to make a mistake is in your classroom. Find creative ways to turn them into learning experiences. If you must, make the mistake yourself and allow the students to correct it. No one wants to see their product used improperly, but don't let that cloud your effectiveness as an instructor.

Capitalize on Informal Learning

The more experience an adult learner has in a broad field, the more they are going to learn in informal ways. They are going to observe the way others do things and let it influence the way they do it. They are going to pick up the tiniest of influences from the most unlikely sources. Humans are always learning. We are always experiencing new things. Some do a better job of categorizing them and pulling them from their memory, but all are learning new facts or reinforcing old ones.

Informal learning can be even more valuable than formal learning, because so much of it is experiential in nature. On-the-job training is usually informal learning. If it is true that it takes on-the-job experience or practice to really become proficient in something, then informal learning is extremely important. Learning in small increments is much easier for most of us than trying to learn everything at once. I'll never forget my father teaching me a valuable lesson when he asked me, "If someone were to offer you the choice between a million dollars now or one penny doubled each day for 30 days, which would you take?" Of course, a million is big number to a child and a penny is a small number, so I chose the larger of the two: one million dollars. Had the offer been real, I would have missed out on more than four million dollars! A penny doubled each day for 30 days equals $5,368,709.12. Apply this simple illustration to learning. Just because your students haven't been through the formal training (offered a million dollars!), they may have gained equally as much or more in small increments.

Your job as a technical instructor is to capitalize on that learning. Don't dismiss or marginalize it. If it needs to be corrected, feel free to do so, but always remember that the pennies they have earned are going to amount to more than your single learning event. Correct it respectfully or it will be rejected.

Allow Students to Share Their Experiences

A typical classroom is full of knowledge. With so much expertise (real or assumed) in a classroom, it is unwise for even the smartest of instructors to pretend to be *the* guru on any subject. One great way to involve students experientially is to let them explain things to other students. This accomplishes two things. First, they learn by explaining. You have probably experienced firsthand that when you teach someone something, you learn it better yourself. Teaching it forced you to articulate your knowledge and helped to solidify it in your own mind. Allow your students to experience that. Second, the other students profit from another source of learning. If you have done much instructing, you have probably had the following occurrence. You made a statement or emphasized a particular fact or thought process. Minutes or hours later a student repeated what you said—either word for word or with their slightly different slant to it. Maybe they didn't even realize they were repeating you. All of a sudden, the proverbial light bulbs seemed to turn on in the other students' eyes. What just happened? How did they understand so quickly from a peer but could not learn the idea from you?

What is happening in these circumstances is the principle of learning from multiple sources. This is a good thing. Next time that happens—and it is sure to happen again—you can either try your hardest to make sure that they understand that you've been *trying* to teach that concept, or you can accept it as part of the learning process. My advice? Thank the student that just "taught" the other student and move on. In fact, you can create these scenarios by allowing students to answer questions before you do. A good trainer cares less about what they teach and more about what the students learn.

Give Lecture and Observation Their Rightful Place

Your students can, of course, learn some things by observation. Lectures, and other forms of teaching, can be helpful for your students. General knowledge may be important, or even mandatory, prior to hands-on application. Learning is not something you can put into a neat pyramid and claim dogmatically how much someone will learn or how much they will retain when taught in different ways. What we do know is that your students will retain more when you let them do what it is you are teaching them. Isn't that your goal?

A trainer needs to be a good speaker because instruction often includes lecture. Lecture is not the enemy of learning. Lecture and observation are just a couple of several means to the end of learning. What your students observe you perform or listen to you talk about is only an inspiration for them to go out and *do* something. If you can bring that "do" into the classroom, by all means, do it.

Provide a Structure for Your Hands-On Training

Hands-on training is more than just a buzzword in product training. It is essential to experiential learning. However, experiential learning needs a structure to increase its effectiveness and ensure its consistency. There are three main phases to hands-on training, each with its own varying grades of application. All three stages are important to produce effective proficiency on a technical product.

Phase One: Exhibit the Product

The first phase of hands-on training is the exhibition phase (Table 2.1). This is when you present your product to the students and they get to physically touch and feel the equipment. This is literally hands *on*. Anyone who offers hands-on training gets at least this far. Most don't offer it soon enough. Get to this phase as quickly as you can in your training. Don't wait for the mandatory 4 hours of slides. You can't examine slides. You can't really tell how big or small an object is. If the equipment is too large to fit in the training room, get to the lab as soon as possible. Don't wait until there is something to do there. Touching the equipment will change the way your students learn about the equipment, so do it as early as possible. If you are training on a software program, find a way to allow them to navigate parts of the software before you start teaching.

Don't forget that this is just the first phase of hands-on learning. In the exhibition phase, the product is the driver. That means that the product size, function, availability,

Table 2.1 Hands-on learning stages.

Approach	Driver	When to introduce	Retention	Response
Exhibit	Product	As early as possible	Short	"I understand it"
Execute	Instructor	After sufficient introduction	Medium	"I can do it"
Explore	Learner	Throughout	Longer	"I did it!"

and so on will determine how the "hands-on" happens. But don't stop there. Occasionally, a trainer will get away with calling his training hands-on because students literally get to see and touch the product they are training on. That's it. Perhaps the equipment is too expensive or too complicated. Maybe they've allowed too many students into the class to actually do anything meaningful with the equipment. But … they can honestly say that they let the students touch the product.

Trainers that are merely exhibiting products call their events "hands-on" much more often than they should. Be honest. You've probably done that yourself. I've had to stand up to more than one manager who wanted me to open up a class to more and more students, but still call the class "hands-on."

Phase Two: Execute a Function

The second usage or phase is when you ask your students to perform a particular function with the equipment. Most honest hands-on training classes offer at least a short execution phase. This phase requires guidance by the instructor or a written guide. The tasks should be specific and practical. The goal is to reinforce the learning of common tasks in a controlled environment.

Execution hands-on training is important. Two things will help make your execution hands-on training more effective. First, let the students execute something on your product as early in the training process as possible. Challenge yourself to reduce or completely eliminate the time between the beginning of class and a hands-on exercise. If you are facilitating a 4-hour class and you were doing hands-on the last hour, see if you can find a way to have them do something with your product during the first hour of training. If you are teaching a 4-day class, and you were doing hands-on in days 3 and 4, see if you can get them to apply something during the first day. Second, take the "hands-on" further. Make it a step toward exploring. Your goal should be to let students discover your product—to learn *while* they are doing.

Phase Three: Explore Independently

To discover one must explore. This phase is less about following specific steps and more about exploring and discovering how it works on your own. This is learning while doing. It is constructing the knowledge for yourself. This is my father leaning over my shoulder as I changed my first tire. He didn't do it for me. "If you want to drive, you have to learn how to change a tire first." What I wouldn't have given for a PowerPoint presentation on how to change a tire, followed by his experienced demonstration. But I wouldn't have been able to say with any measure of decisiveness that I knew how to change a tire. Substitute your own experience. I'm sure you have one. The point is that a presentation or even a demonstration was unnecessary for my learning and it is often unnecessary for your students as well. One suggestion is to add a troubleshooting element to your training. Elements like troubleshooting force the student to think on their own and can significantly increase the effectiveness of your product training.

Without hands-on training you have only introduced a topic; you have not taught it. If your goal is to introduce your product, you may not need to have hands-on training. But if your goal is to help others become proficient in using your product, you need to provide an opportunity for the student to apply what they have learned. If it isn't possible to use their knowledge in the classroom, allow for a process in the field or on the assembly line that gives them additional learning opportunities. Only after they do whatever it is they are supposed to be able to do with your product can they become proficient in it.

Apply All Three Phases

All three of the hands-on approaches are helpful. One mistake instructors often make is to separate the hands-on portion of training from the lecture. Lecture typically gets priority. By the time students are allowed to touch the equipment, there is not enough time to progress past the examine phase. Some are fortunate to get into the execute phase, but almost none get to explore the equipment. The solution is simple. Don't make your students wait to get their hands on your product. You don't have to. If you were going to show your best friend how your product works, would you start with a presentation or would you hand it to him and say, "Look at this, isn't it cool?" Do the same for your trainees. You'll find that doing this sets the tone for the rest of the training—this training is not going to be death by PowerPoint. It also sets the context—this is the product you are going to learn about. Finally, it sets the expectation—this is what I expect you to be able to do after this training is over.

Conclusion

I have emphasized in this chapter what technical instructors refer to as "hands-on training." I've demonstrated that not all training events that allow for students to handle the equipment are the same. They are, however, related and each deserves a place in your classroom. Be deliberate. Know which one you are using and why.

Throughout this chapter I have maintained that no one can become proficient in something that they haven't done. There is a relationship between what you know and what you do, but doing is what creates understanding. The same is true for you, in regard to teaching. You cannot become a great instructor until you start instructing. Seek out opportunities to teach about a product or technology you are knowledgeable about. See if you can get the students to do more and listen less. Watch their eyes light up as they tell you what they've learned without a lecture or presentation. When that happens, you will sense that your role has shifted from being a public speaker to being a facilitator. You'll find that you worry less about how you deliver the content and more about how the students learn it.

And that's why knowing how adults learn matters to you. Emphasizing experience over knowledge is a skill that requires practice. But when you do, you will find that it leads to an enjoyable, effective learning experience.

Making It Practical

For learning to develop into expertise, there must be experience. Take a moment to consider a technology, product, system, or other technical skill that you are competent in. Think about the role experience played in your learning process.

1) Name a skill you are proficient in.
   a) What is one way you were shown or demonstrated the skill prior to becoming proficient in it?

   b) What is one exercise or guided task you completed as you were learning this skill?

   c) How did you explore this product or technology as you were learning?

2) What will you do differently in your next training session as a result of reading this chapter?

Before you read Chapter 3, "You Know It, Can You Teach It? Overcoming Your Own Intelligence," answer these two questions:

1) Explain the statement "You can't teach what you don't know you know" in your own words.

2) What is an appropriate response when you are asked to teach something, but you are not given enough time to do it?

Note

1 Cambridge Dictionaries Online. http://dictionary.cambridge.org/us/dictionary/english/skill (accessed August 8, 2017).

3

You Know It, Can You Teach It?

Overcoming Your Own Intelligence

Knowledge is power—at least that's what we've all heard from multiple sources and with different applications. Usually, people who want more power are the ones making the statement. Rarely, do you hear someone state "I have knowledge; therefore, I have power." The statement, though not wrong, is incomplete.

Power is an energy source—the source to do something. When energy is misapplied, it can be destructive, or, at best, ineffective. As a preschooler living overseas I remember watching my father painstakingly put together a Hi-Fi stereo system he had brought with him from the United States. After what seemed like hours of attaching the radio, amplifier, and turntable system together, he finally plugged the stereo into the wall outlet. I remember the smell. To his dismay, he had just plugged a 110-V system into a 220-V outlet. More power was definitely not better.

As an expert of your product, you are like the 220-V outlet teaching 110-V students. You need a voltage regulator or transformer to be effective. You need to slow down. You must reduce the content you would like to provide and give your students what is best for them, instead. As you teach, keep in mind that you didn't become a product expert in a day and you aren't going to make anyone else an expert in a day, either. This is not a question of withholding information, but of offering the right information at the right time. This is about learning how to give your students time to create true understanding for themselves.

I have been fortunate in my career to work with very intelligent individuals. Most were competent and valuable employees. I have found very few engineers to be "knowledge hoarders" who fear power sharing with those they enlighten. To the contrary, I find that most love to share their knowledge in a coaching-type environment, though some are not as effective in front of a classroom. This problem is not unique to engineers or technical experts. Universities are full of professors who rank among the top experts in their fields but are ineffective teachers.

Address Your Biggest Challenge: Yourself

For a variety of reasons, many experts struggle to "transfer their knowledge"—or to help others construct the knowledge that they already possess—in a classroom setting. Some of those external reasons will be addressed later. The biggest challenge that you, like any expert, will face is yourself. You must overcome your own natural abilities with your product.

Product Training for the Technical Expert: The Art of Developing and Delivering Hands-On Learning,
First Edition. Daniel W. Bixby.
© 2018 John Wiley & Sons Ltd. Published 2018 by John Wiley & Sons Ltd.
Companion website: www.wiley.com/go/Bixby

Fortunately, overcoming your own expertise is not an impossible task. Managing your own intelligence is only daunting when you don't know you need to or are unwilling to change the way you teach. Intelligence alone does not make a good instructor. Ability can be a terrible teacher. Both are important. But both often get in your way.

When you teach, you will unconsciously put yourself in the place of your students. Your tendency will be to teach them the way you want someone to teach you. That type of teaching does not work unless one has already achieved expertise. Your students cannot learn in the same way that you can. Good instructors are not less intelligent, but they are able to understand their students and communicate at their level.

The Four Stages of Competency Applied to Instructors

Several decades ago, Noel Burch developed a model for skill development that was quickly adapted by leading psychologists and educators. The model is brilliant in its simplicity and is helpful for experts, not only to understand their student's level of competency, or skill, but also—and even more importantly—to understand themselves, and why they have a hard time teaching.

You can best understand the model as an XY graph, with "Consciousness" on one axis and "Competence" on the other, as in Figure 3.1. Competence starts in the lower left side of the graph and moves to the lower right side of the graph. This model represents the progression from being completely unaware and unskilled to being a true expert in the field.

None of these levels represent aptitude, in either a physical or a mental sense. This is not a general progression from ignorance to intelligence, but rather a very specific application of very specific skills. Any individual will find themselves at level 1 for some skills and at level 4 for others.

Figure 3.1 Four stages of competency.

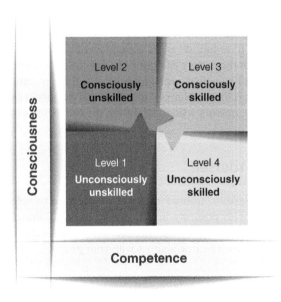

Unconsciously Unskilled

This level, also known as *unconscious incompetence*, is where we all begin. At this level, an individual is unaware of what they even need to learn to perform a skill. They may not even know the skill exists. You may have heard the overused phrase "they don't know what they don't know." When applied to a skill, the phrase refers to those in the first level of competency.

It is possible to get students in a product training class who are unconsciously unskilled. They may be coming with no idea about what your product does or that it even exists. Perhaps a manager sent them to your class, or they just happened upon it at a trade show. The good news is that the easiest and quickest transition is from stage 1 to stage 2, *consciously unskilled*.

> I know submarines exist. In fact, I have toured one. However, I don't even know if you drive one or if you pilot one. I don't know enough about submarines to know what I would need to learn or where to start. I am definitely unconsciously unskilled.

Consciously Unskilled

The second level is also known as *conscious incompetence*. At this level, the student has moved from "not knowing what they don't know" to "knowing that they don't know it." Hopefully, most of your students will be at this phase. They know that they need to learn something about your product and have a desire to increase their skills.

> I love watching a skilled excavator work with a backhoe. I have operated a backhoe simulator. I know that it operates by pulling levers, pushing pedals, and moving around a joystick. However, you do not want to hire me to dig a ditch or basement. I have not practiced it in real life, and I know that I am not good at it.

Consciously Skilled

The third level is also known as *conscious competence*. At this stage, an individual has acquired a skill and "knows what they know." Generally, they also know how they learned it as well. Most trainers will see this stage as a developmental or continuous learning stage, since the individual has not mastered the skill to the point that it becomes easy for them.

> I enjoy flying kites. I like the challenge of getting it in the air and keeping it there. When they were young, my children thought I was an expert at it. Unfortunately, however, I rarely fly kites, and when I do, I have to think through everything I'm doing to make sure to get it right. I usually do, but it takes some thought on my part.

Unconsciously Skilled

The fourth and final stage is also known as *unconscious competence*. At this stage, an individual demonstrates the highest level of skill—so much so that it has become almost second nature to them. At this level, one doesn't even have to think about what they are doing; they have done it so much. You might relate it to muscle memory or reactive instincts.

> For over 25 years, I have driven a car. I am an unconsciously skilled driver, regardless of what my wife says. I don't have to think about where the brake is or which way to turn the steering wheel when I want to turn. I can comfortably hold a conversation and drive at the same time.

Why Experts Find It Difficult to Teach

Generally, employers and educators alike reward the fourth level of competence as the ultimate achievement. This is where you want your skills to be in order to be a great employee. If what we do is truly what makes us valuable, then doing it at the highest level makes us the most valuable.

However, while doing something so well that it has become second nature to you definitely proves your ability to perform that task, it makes it impossible to teach. You cannot teach an unconscious skill. Said differently, you cannot teach what you don't know you know. This is how your own "intelligence" can make you an ineffective instructor. It is not because you are smart, but because you are so skilled that what you do is easy or habitual for you. This is the reason it is so rare that sports superstars make good coaches. Their ability and skill becomes so second nature to them that they cannot explain to others what they need to do to replicate it.

There are at least three reasons why you can't teach something you are very good at. First, you don't know how you learned it. Second, you can't distinguish between the easy and the difficult. Third, you can't differentiate between the essential and the nonessential.

Experts Rarely Remember How They Perfected Their Skill

When you do something extremely well, you have systematically, and perhaps, deliberately de-emphasized or forgotten certain things. One of those things is how you learned it. I don't mean that you don't remember your first training class or a particular lesson. But those were just events in a long learning process. Over time, you have added together many experiences that have accumulated to make you an expert. You have made many mistakes and you have learned from them. You have tried things—some that worked and some that didn't. You have gathered information in bits and pieces and don't remember where or how you learned certain things.

You also don't remember the sequence in which you learned them. You don't know which items you learned first and what was simply added on through experience. If your boss asks you to teach a class today, you will likely start with what is closest in your memory even though what is furthest in your memory may be the proper place to start teaching.

Experts Have Trouble Distinguishing Between the Simple and the Difficult

When you are an expert, everything is easy. Easy, of course, doesn't mean it takes no physical or mental exertion. It doesn't mean it takes no extreme focus or attention to detail. It simply means that the process has become normal for you. Easy is also relative. If you can do something that I can't do, regardless of how much effort it takes you, it is easier for you than for me. You can do it, but I can't. It isn't easy for a surgeon to insert and control a guide wire through a vein and perform microscopic surgery. It isn't easy for a lineman to climb to the top of a tower and connect a communication cable, or an engineer to make a computer-assisted drawing of someone else's latest idea. None of those are easy, but when done by a skilled person, it looks easy to the casual observer.

Before becoming an expert, you had to first master the difficult. You intentionally made the hard things look the same as the easy things. If only the easy things are easy for you, you are not an expert. It is this lengthy process of equalizing the simple and the difficult that made you valuable. It is exactly that process that must be undone to make you a good instructor.

Experts Don't Differentiate Between the Essential and the Nonessential

Another line you have blurred over time as an expert is the differentiation between the essential and the nonessential. I'm not referring to the extremes. Usually, there are things on both ends of the spectrum that are obvious, even to the beginner. There are certain things that are extremely essential and others that are obviously not essential. I am referring to the bulk of learning between those extremes. If 10% is obviously essential, and 10% is obviously not essential, that means that 80% of your curriculum could fall into either category. Most experts have a difficult time prioritizing that 80%.[1]

The blurring of these lines, however, occurs differently than with the difficult versus the easy. Often, the addition of nonessentials is what turned your skill into an art. It is the way *you* do it. Nonessentials can distinguish you from even other experts. You have perfected them over time and they are your signature on your industry or product.

The problem with emphasizing your signature is that when the process is unique to you, the nonessential has become very essential to you. Good instructors, however, know the difference between what is essential and what is not. Good instructors are willing to set aside their preferences to teach the basics. There will come a time to teach others your nonessential signature. But first, you must be able to teach the basics.

How Experts Can Teach It

The great news is that experts can make remarkably effective instructors. It is preferable, in fact, to have an instructor that has reached the highest level of competence. No one wants to learn from a novice, and no novice wants to teach experts. The only way you can become a good instructor, however, is to force yourself to go back to the consciously skilled stage. You cannot teach at the fourth level of competency. Going back to the third level is a skill—one that, over time, you can perfect. That is why instructors should teach often. You can become unconsciously skilled at moving back to the consciously skilled level, but it takes practice (Figure 3.2).

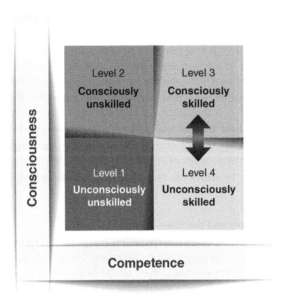

Figure 3.2 How experts use the four stages of competency to teach.

Ask the Instructor (Yourself) the Right Questions

Good students ask good questions to the instructor. That is how students learn and improve. In the same way, good instructors ask good questions to the instructor (themselves). Here are some questions you should ask yourself whenever you face the need to move back to level 3 competency:

How did I learn that? Asking yourself how you learned something forces you to connect the dots between formal and informal learning. Put yourself back in time as a student and force yourself to "learn" it all over again. Are there steps you had forgotten that you had to learn? Are there concepts that are important that you might have missed, or ones you thought were important that you didn't learn until later in your career, after you were already an expert? Maybe those are still important, but at least it puts them in perspective.

What am I doing? It is surprising how many experts cannot answer this simple question. They don't know what they are doing or how they are doing it. Of course, they know the big picture, but it has been so long since they've thought through all the little steps that they don't even know what they are really doing. It is your ability to do the little things effectively and efficiently that make you an expert. It is being able to explain those little things that will make you a good instructor.

I remember having to do this when attempting to teach my daughter how to drive a manual transmission vehicle. It is second nature to me. I couldn't teach her until I deliberately went through all the steps again myself. Once I had them down, I could explain it to her. Until then, I only frustrated her by making it appear so easy.

What am I assuming? Making assumptions when you are teaching is normal. You have to. You must also know what those assumptions are. I find that simply stating the assumptions to myself helps to clarify my objectives. If both beginners and advanced students make up your audience, it will take some creativity to keep them both interested. Regardless, you need to eliminate the easy assumptions. For example, don't use acronyms that the students may not know, or jargon particular to your industry or company. If a level of expertise is expected, be sure to state that clearly in advance. If it isn't, be sure to start the learning at a point where even beginners can start the learning process.

What if they only learn one thing? For a subject matter expert, this question is one of the most difficult to answer. Do not answer the question with vague words like "an overview..." or "a general understanding of...." Instead, narrow it down to one thing. Answering this question specifically will demonstrate an understanding of how to teach. Knowing this answer will help you be more successful ("Hey, they learned the one thing I wanted to teach them!") and it will help to determine what really is essential and what is not.

Conclusion

To become an expert instructor, you must first understand that teaching is a separate skill that requires a new way of thinking. Most importantly, like any other skill, teaching requires practice. You must do it. Your expertise can become a significant advantage to you as an instructor, but you must develop the separate skill of making that happen. It is true that teaching comes naturally to some, just as almost every other skill comes naturally to some. The gift is a rare one, however, and much more so in the field of technology. Even if it is, you must challenge yourself to overcome your own familiarity with your product.

Think about the task of cutting out a figure on paper. It is possible to cut it out with a sharp knife. But the job is much more effective when done with a pair of scissors. The scissors are really just two sharp knives working together. Your challenge is to take one sharp knife, the expert talent you have with your product, and pair it with another sharp knife, an expert talent in teaching. That is when you will become a great instructor.

Making It Practical

All adults are skilled in some things and unaware that other skills even exist. The more proficient you are in any skill, the more difficult it can be to explain that skill well to someone with no proficiency or experience. Understanding the theory behind that truth can be helpful in becoming an effective instructor.

1) What level of competency (skill) makes the best instructor?

2) What level of competency (skill) makes the best student?

3) In your next training class, how will you emphasize the essential and de-emphasize the nonessential?

Before you read Chapter 4, "Ready or Not? Why Some Students Are More Ready to Learn Than Others," answer these two questions.

1) Why do you think some students learn more than others in the classes you teach?

2) Have you ever attended a class you weren't interested in but ended up learning a lot from it? Why did that happen?

Note

1 The numbers I'm using here are for discussion purposes only. They do not indicate or reference any study or empirical evidence.

4

Ready or Not?

Why Some Students Are More Ready to Learn Than Others

As a subject matter expert, you are likely excited to teach new students about your product. It can be frustrating, though, to find that some are not as eager to learn as you are to teach. It can also be overwhelming when they are overly enthusiastic about learning and you can't teach them fast enough.

Why is it that some students are more primed and ready to learn than others? How can you say something to two equally bright students and it will change one's behavior and knowledge and have no effect on the other? It is important to understand what drives a student's readiness to learn and what you can do to influence them.

There are, of course, external influences that are beyond your control. Any number of external influences can make learning challenging, even for bright students. Emotional or physical stress can be a barrier to learning. Lack of sleep, for example, may make learning difficult for a student. While you should make every effort to schedule the classes to reduce this problem, the reality is that you have little control over your students' sleep habits. You will have even less control over the emotional stress that may be hindering their learning.

The Four Principles of Learner-Readiness

Being aware of your students' physical and emotional state is important. A good instructor will adjust schedules and even learning requirements accordingly. However, your concern is primarily with what you can either control or influence. There are four principles that must be true of a student if he or she is going to learn in your class. A student who meets all four of these requirements will learn best, while one who is missing one or more will struggle to internalize the learning. The four learner-readiness principles are that learners must recognize the need for learning, they must take responsibility for their learning, they must be able to relate it to past experiences, and they must be ready to apply the learning.

Product Training for the Technical Expert: The Art of Developing and Delivering Hands-On Learning, First Edition. Daniel W. Bixby.
© 2018 John Wiley & Sons Ltd. Published 2018 by John Wiley & Sons Ltd.
Companion website: www.wiley.com/go/Bixby

They Must Recognize the Need for Learning

> Why do I need to learn it?

Because learning is personal, the motivation to learn is personal. One of the questions students always ask in any structured learning environment is "So what?" They probably don't vocalize the question, but they want to know how the learning will benefit them.

In a typical product training class, you will have some students who are there because they want to learn. Those students have already answered the "so what" question for themselves. Anyone who wants to learn something has an idea of how it is going to benefit them to do so. You may also have students who have yet to answer that question. It is important to help them do that as soon as possible.

There are many ways to determine if your students have answered that question or not. One way is simply to ask. At the beginning of a class, ask students to state one thing they hope to get out of the class. Ask them why they are attending the class or what they hope to do better after the class. Ask for specific answers. If you get the typical "to get a better understanding of your product" answer, try to dig a little deeper. Ask how that "better understanding" is going to help them or their company. Doing this very early on is important, since the student must answer this question before learning that will last can begin.

What if Their Reason for Learning Is Wrong?

Sometimes as you're asking that question, you'll get someone with an obvious misunderstanding about what your product does or how it will be beneficial to them. Depending on the circumstances, this can get a little tricky to navigate, since you don't want to embarrass a student in front of the class. You want to minimize their misunderstanding as much as possible and correct it as soon as you can, preferably in private during one of the first breaks.

If your product is particularly innovative or new to the market, it is more susceptible to a misunderstanding of its benefits early on. Asking leading questions in these circumstances can be helpful to eliminate public misunderstandings. Another way is to encourage some type of pre-learning before class begins. A short video or quick guide can help students answer the "so what" question before coming to class.

They Must Take Responsibility for Their Learning

> Will I put forth the effort to learn it?

The second principle of learner-readiness is that the student must be willing and ready to take responsibility for their learning. Understanding is earned. Learners must own their learning. Challenge your students to review the stated objectives and choose one that is the most important to them—which one they want to be certain to learn during the class. This not only helps them recognize their need for training but also helps them to take responsibility for it as well.

But not all students understand that they must take ownership of their own learning in order to change behavior through education. Most assume that they will learn proportionately to the value of the content or the ability of the instructor to teach it. Both are, of course, important. Adding the third pillar of personal responsibility is often something they aren't used to hearing in a corporate or technical learning environment and has the potential to make a significant difference in your class.

Questions Demonstrate Learning

One way to encourage responsible learning is to teach your students how to ask questions while they are learning. Don't wait until the end of a dialogue to ask questions. Students often won't ask questions then, unless the question is about something you covered toward the end of your presentation. If you encourage asking questions throughout the class, it will help increase their level of responsibility.

The right reason to ask questions needs to be encouraged as well. There are many students who believe questions are merely the way they can clarify something they didn't understand. If that is their understanding, then the need to ask questions is a result of one of two options: either the instructor didn't explain it well, or the student was the only one unable to absorb the information the instructor gave.

In order to avoid looking unintelligent, many students will assume that the instructor didn't explain the concept well. That allows them to ask the question without looking ignorant in front of their classmates. Those students usually ask for clarification for everyone. They are often not satisfied unless you explain the entire concept in a new and fresh way, which can take significant time.

Asking questions, however, is really a tool that learners use to build new information on previous experiences and learning. Since all previous learning and experiences are unique, it has nothing to do with intelligence and may have little to do with your explanation. Good questions have everything to do with taking ownership of the learning.

Encourage your students to ask questions. Create a positive motive for asking questions by changing it from lack of intelligence to a desire for more intelligence. Make it a positive experience when they do ask a question. Thank them for helping to clarify something for everyone and make sure they understand it before you proceed with something else. In exceptional cases, when it is obvious that they don't have the experience to use as a foundation for new learning, ask for time outside of the classroom to help them individually. That is the spirit of a great instructor.

The Instructor's Responsibility

While students must take ownership of their learning, instructors have a responsibility to help the students learn. When a sports team loses a closely contended game, it can be difficult to determine if the failure to win should rest on the coach's shoulders or on the players'. Often it is both. While the players must execute and take ownership of it happening, the coach is responsible to make it possible.

The same is true in the learning environment. You are a coach. Your main job is to provide them the right opportunity or environment in which to learn. You cannot learn for them, but you are just as much a winner as they are when they do. Like a coach, you need to be continuously evaluating your students, perhaps even adjusting your strategy to make sure a change in behavior actually occurs. Stop teaching new things if the previous concepts haven't been learned yet. There is no value in moving forward until your students are ready to learn.

They Must Relate It to Their Experience

Can I relate it to what I already know?

Since all learning is a process of building on previous learning and experiences, you cannot simply provide facts and expect students to retain any new learning if they can't relate it to their past knowledge. Students must be able to relate the new learning to learning they have already acquired or experienced.

One of the best ways for subject matter experts to help students relate to the learning is to relate it to their own experience. Many experts think that stories and anecdotes are time-wasters that have no real learning purpose. When used correctly, however, they can have a powerful influence on facilitating learning.

Tips on using personal experiences:

Tell about your own learning experiences. When you tell a story about your past learning experiences make sure it is just that—about learning. Never tell a story to try to impress your students with how well you used your knowledge in a particular circumstance. That may boost your ego, but it will have little effect on their learning. Instead, tell about when you learned or applied the concept and what you learned from it.

Make yourself vulnerable. Don't pretend to have always been a natural if you were (or are) not. Tell them about mistakes you have made and what you learned from them. Being vulnerable will not diminish your students' respect for you, but that is not your main concern anyway. Your main concern is for them to learn. When they sense that you are trying to help them avoid some of the same mistakes you have made, they will be grateful for your leadership.

Use recent examples. If possible, use examples that are fresh and recent. It demonstrates that the learning never ends, even with an expert.

Keep your examples relevant. If there is not a good objective for telling the story, don't tell it. Recounting experiences can be a powerful way to help students learn, but as soon as it becomes irrelevant, it dilutes all future stories you may want to tell.

Keep your examples short. Yes. That short.

Know your audience. Unlike your learning objectives or your outline, you can have multiple examples and use different ones every time you teach. You should know your audience well enough to know what will most benefit them. Be culturally and educationally sensitive. Emphasize their learning, not your amazing experience.

Listen to their stories. One of the hardest things for an instructor to do is to encourage others to tell about their learning experiences. It can be a challenge to keep them relevant, recent, and short, but when done well it can be a great learning experience for everyone. Never try to compete with their story. Instead, be grateful for the learning experience. Demonstrate what you can learn from their experiences to the class. This will help them see how to take another person's learning experience (including yours) and apply it to their own circumstances. Hearing stories from multiple sources increases the probability of being able to relate the learning to their own situation.

They Must Be Ready to Apply It

> When will I need to apply it?

Another question students are subconsciously asking themselves when they come to class is, "When am I going to need this?" It is the facilitator's job to get them ready to apply their learning, and students get ready faster when they believe they need to apply the learning soon.

The principle of proximity to the need is what makes on-demand training so powerful. The immediate need is what makes it effective. My son and I recently installed a new sound system in his car. We watched several YouTube videos prior to starting the project and several more when we ran into issues putting the dashboard back together. Those videos, and many others, proved very helpful when I needed information. However, if I just sat down to watch the videos for entertainment purposes, I likely would not learn much from them. Not having the immediate need for the learning removes the value almost completely.

The same is true of your learning. If they don't ever think they'll use it, they are likely to learn very little. Even knowing they will use it but not having a specific need in the near future puts them at a significant disadvantage. If your students know the need, build on it. If they don't, you need to create it.

One of the values of hands-on training is that it creates that need. When students know that they are going to have to perform the task you are teaching them, they will be more eager to learn it. This is more than just a desire to pass a test or save face in front of the rest of the class. This is an actual learning principle. If you show a student how to do something just *in case* they ever need to do it, and then you show them how to do something because they *are* going to do it, they will always learn better in the second scenario.

Conclusion

Product solution training can mean teaching a wide range of students, from a disinterested employee to a highly engaged consumer. It can be challenging when one student learns quickly and is absorbing the learning as fast as you can teach it, while another one is being left behind. The first thing you can do as an instructor is to understand the reasons why this happens. The second thing you can do is find ways to influence your students. If he or she isn't ready to learn, you're not ready to teach. Don't begin proficiency training without preparing them first.

Most students come into a technical training class with an obvious understanding of their need for the training. If they don't, it is wise to offer short presentations that focus on the need to learn more and how that will help the student. You can make them part of your hands-on training, or you can deliver them as a prerequisite learning module.

Your influence over a student's readiness to apply the learning, their efforts to take responsibility for their learning, and their ability to relate it to their own experiences will vary depending on your audience and the solution you are teaching. You can influence all of them by engaging your students in the learning experience. As you engage them in learning exercises, you will be encouraging them to take responsibility for their learning. As you engage them in questions and conversations, you will be helping them to relate to what you are teaching them. As you engage them in hands-on labs, you will be creating a readiness to apply it, even in the classroom.

Your students' readiness to learn is influenced by a DO versus KNOW philosophy of learning. Not only do adults learn best when they are doing, but also they are most motivated to learn when they are doing. Get your students motivated to learn and they will be easy to teach.

Making It Practical

Perhaps the most important job of any instructor is to motivate his or her students to learn. Take some time to reflect on the four learner-readiness principles: (i) students must recognize the need for learning, (ii) students must take responsibility for their learning, (iii) students must relate the learning to their previous experiences, and (iv) students must be ready to apply it.

1) Which do you feel you have the least control of?

2) How are you going to change that in your next class?

3) What is one way to get your students to take responsibility of their learning?

Review of Part One: The Foundation of Hands-On Learning

1) What is one thing from Chapters 1 to 4 that you will apply the next time you teach?

Part II

The Strategy of Hands-On Learning

5

It is Never Just Product Training

Why You Should Offer the Training

There are many reasons and ways to educate adults. Corporate training programs come in many shapes and sizes. They don't all have the same approach, nor do they all have the same expected outcomes. Some educational programs exist to improve or build on a student's soft skills, while others seek to teach new proficiencies in a technical or precise environment. Some require extensive proof, not only of comprehension but also of the ability to perform the required task correctly and consistently. Some training programs require only comprehension testing or no testing at all. Others exist purely to avoid a lawsuit. I am sure you've had to sit through a few of those or at least click through the online compliance modules while you tried to do something more "meaningful."

Applying the adult learning principles that were covered in the first few chapters will make you more effective as a product instructor. What it will not do is ensure that the training helps your business—and training that doesn't help your business isn't worth delivering. Offering training for the right reasons is essential. Otherwise, your most effective delivery will have ineffective results.

Product Solution Training Versus Talent Development

It is critical to the success of your product that you be effective as an instructor. Unlike some of the training mentioned earlier, you are teaching about a product or solution that you produce. Without competent users, your product has no purpose. Creating competent users is not just important or helpful; it is essential.

Unfortunately, many companies do not put a strong enough emphasis on product training. It's not that they think product training is unimportant. It is more likely that they are caught in a risk–reward cycle that prohibits them from investing too much in product education. Corporations are reluctant to spend money measuring the results of a program they have little investment in to begin with. Those that do find the investment in their customers' talent development to be similar to the investment in their employees' development: extremely beneficial.

Product Training for the Technical Expert: The Art of Developing and Delivering Hands-On Learning, First Edition. Daniel W. Bixby.
© 2018 John Wiley & Sons Ltd. Published 2018 by John Wiley & Sons Ltd.
Companion website: www.wiley.com/go/Bixby

Whether it's because it is the right thing to do, because they realize it works, or because it is trendy, most companies invest significant amounts of money each year in developing the talent of their own workforce. Associations and magazines, conferences, and seminars all join to make talent development a multibillion-dollar industry.[1] The more companies invest in talent development, the more important and valuable the results are to them. As they invest more, they realize how effective the development can be. They build intricate metric models to fine-tune their offerings. They don't just offer training classes; they want to make sure that the training they are offering changes (or *develops*) the talent of the student. Developing talent is a significant part of the strategy of most successful business models.

The sad reality is, however, that many of them do this so that they can offer a product that is superior and more efficiently produced than their competition but then neglect to spend that same energy to develop the talent of those actually *using* their product. According to most studies, only a small percentage of money spent on training goes toward product training, and the great majority of that is dedicated to employee product training.

Employee Product Training

The result of putting such a small emphasis on product training is that corporate product trainers are at a significant disadvantage. Proven, effective training courses are not readily available on your products. Employees are expected to get most of their product training on the job. This makes the learning less consistent and makes metrics or proof of success virtually nonexistent.

When formal training on the product does exist, it is often delivered by you, the subject matter expert. You may have even had to develop the curriculum, which likely meant you or another subject matter expert put together content that contained everything about the product that you thought was necessary for the student to learn.

By making product training the responsibility of those who know the most about the product, the company is recognizing the importance of getting the subject matter experts involved. It could be dangerous, and potentially even irresponsible, to put the training in the hands of those who don't know enough about the product. What most people are missing, however, is that transferring that knowledge requires a completely different skillset. Ignoring that may be equally irresponsible.

I regularly hear statements from corporate leaders that confirm a gap between product training and talent development. In a recent project, I asked if there was any intention of tracking the requested training within the human resources talent development program. The reply was simple: "No, this is just product training."

Except that it isn't. It's never *just* product training. Even when the target audience is internal employees, the product training they get should be the best in the industry. If you make a unique product, and you are the only source for that training, don't let that become an excuse for subpar training. Your company will only benefit from training that changes behavior, and that should be the foundation of your business strategy.

Customer Product Training

When you are asked to put that same training together for customers, your challenge is even greater. Now, your students are from outside of the company. You don't have a captive audience, nor do you have access to the students' employment or previous training records. It is more difficult to follow up after a training event to verify that the training is, in fact, changing behavior. Even if you wanted to tie it in with their talent development requirements, it would likely be impossible.

The good news is that greater challenges bring greater rewards. You now get to change what customers do with your product or solution, which can bring greater returns to your company and product. You should do so using many of the same metrics, the same learning principles, and the same management philosophies that your counterparts in talent development are using.

Because the talent development field is large and profitable, there is a lot of competition ready to help your company develop that type of training. Competition can help to breed excellence. Unfortunately, as an expert that trains on your own product, you likely do not have much competition, if any at all. You may have put together the content yourself or relied heavily on another subject matter expert to provide the training content for you. This is especially true if yours is a start-up company or one that produces proprietary products. Everyone that takes your training thinks it is great. They have to. No one else offers it, so it is the best there is! It is easy to get flattered by great student evaluations, but difficult to improve and know for sure that you are offering great education on your product or solution.

Good training, when paired with a good business strategy, can be the lifeblood of a product or an entire company. Good training that isn't backed by an effective strategy, however, can be nearly as bad as bad training, which is catastrophic to a product, process, or program.

As a technical expert, you probably have, or have had, other job responsibilities in the company. You are aware of all the work required to develop and produce your particular product. It stands to reason, then, that you, as much as anyone, do not want the training on your product to be developed with any less care or importance.

Business Plan

Creating a solid business plan is the first pillar of effective learning. It is your strategy. The first element to understand in your strategy is the business reason for offering the training. Usually, businesses take one of two approaches. Either training is viewed as a necessary service—a cost of doing business, or it is seen as a profit center that must make money, based on the intrinsic value of the training courses themselves. There are pros and cons to both of these approaches, but I will suggest a third, more business-friendly view.

Training as a Cost of Doing Business

When companies develop product training as a cost of doing business, there is usually no initiative to sell the training or even an attempt to recuperate costs. Most salespeople prefer to see training as a cost of doing business, and chances are, if your training department sits in your sales organization, this is the preferred approach. Many salespeople are afraid of anything interfering with a potential sale. All trainers are part salesperson and don't want a lack of knowledge to hinder the purchase of a product.

This approach can be effective. When a product is new to the marketplace or there is little demand for it, offering training as a free perk is often a necessary way to create a demand for training. Ease-of-use issues or quality misperceptions due to lack of knowledge about the product might also force you to offer your training free of charge. Salespeople can be the training program's best advocate in these situations, and training leaders do well to listen to them. Salespeople are right to insist on simplifying the purchase and use of your products. While I have had more than one executive telling me emphatically that they are not going to charge their customers to learn how to use their product, the reality is that they do. They charge the customer to own the product in the first place. So they are willing to charge them to use it, but not to learn how to use it.

In the cost of doing business model, training is no different from the packaging in which the product is delivered. The idea is that the cost of educating the customer is built into the cost of the product. The reason for not charging makes sense. If you don't charge for telling them about the product before they buy it, why should you charge to tell them about the product after they buy it?

There is always some opportunity to offer training as a cost of doing business. Sometimes it is the only way to get training to the customer and must remain an option to training leaders. However, it should rarely be the only business model for your training department.

Training as a Profit Center

Another favored approach to customer education is turning the training into a profit center. Most often, those companies taking this approach sell more than just a simple gadget or stand-alone product. They often sell a solution that must be integrated into a larger system, where training is as important for the success of the buyer as for the seller. This creates an opportunity to treat training as another product that can increase profits.

Running a training department as a profit center is not an easy task. It must be a very lean department and often the delivery method trumps the quality of the training. The first question asked is rarely "what is the best way to educate my customer?" but is usually "what is the least costly way to educate my customer?"

The increased availability of eLearning has certainly helped to make training more profitable. There are many positives about good eLearning beyond the cost savings. Making training available for students when they need it most is a real possibility with eLearning and is even more accessible through mLearning (mobile learning). It is not always the most effective approach, however, and many product trainers are afraid to voice their concerns about using these methods exclusively. This is especially true if their executives have put pressure on them to use technology to save money.

Two issues are very common. First, executives tend to see eLearning as the cure-all approach. Second, success is driven by the wrong metrics—a word on both of these issues.

Caution 1: Technology is not a cure-all. eLearning, as wonderful as it is, is not the best way to teach everything. When the modality of the education becomes more important than actual learning, your training may cease to be effective. Just as content is not king, neither is the delivery method. You must start with the desired business and behavioral outcomes in mind before determining the modality of the training.

In order to develop effective eLearning, corporate leaders must make a significant investment. The natural desire to get the most out of that investment forces some executives to emphasize eLearning technology over effectiveness. If they fail to understand the concept of proficiency training as opposed to product presentations, they may accept the mere delivery of content as success.

Caution 2: Success may not be success. This leads to the second issue that can arise when tasked with running a profit center: the wrong metrics will almost always determine success or failure. Sometimes training programs are called a success when, in fact, they are not increasing sales of their company's products. When you sell training, you need people—anybody—to take the training. It is as simple as that. It doesn't matter that the right people are taking the training, or even that the training is successful. As long as you can get a lot of people to purchase the education you claim to provide, and you can keep the cost of producing that training to less than you are selling it for, you will be successful. Except that you won't be. Real success only happens when you generate more product sales. Even the goal of reducing costs to deploy or use your products, or the goal of creating a valuable service for your customers, is really part of an ultimate goal to increase sales.

It is also true that some training departments that are generating sales of a product have been shut down due to failure to meet profit demands. This happens when companies don't provide a way to link sales to product training or when product trainers are unable to demonstrate the connection.

With the exception of a few very large customer education programs, most of those that call themselves a profit center are not, in the true sense of the word, "profit centers." Instead, charging for training is usually an attempt to cover the expenses of what they perceive to be a necessary service.

But making a profit is good! Just as it is true of running the department as a cost of doing business, so it is true that there are times when it makes sense to run the customer or product education department as a for-profit center. If your product is first or second in the market or there is an external demand for training on your product (such as end-user or regulatory requirements), it is reasonable to run the department as a profit center.

Charging for the training is also good. Don't confuse putting a price on the training with a requirement to make a profit. Too much training is given away. The result is the devaluing of the education to both the company offering the training and the students taking the training.

So there are positives and negatives to this approach as well. The positives are that it builds a culture that assigns a value to the learning received. It also forces a lean and efficiently run department. The negatives are that it emphasizes the wrong thing, encouraging the wrong metrics to determine success. It often means that the training is designed out of house. Many times the trainers are not as qualified as they could be. In an effort to keep costs down, the training is spread too thinly across many product experts and the skill is never fully developed. In this model training managers are often industry experts, not adult education specialists.

Training that Sells Products

This leads to a third way—a blended approach to managing a training department. Educating users of your products is neither a cost of doing business nor is it the main focus of your business. Of course, I am not referring to companies for whom providing training *is* their main focus nor am I ignorant of the exceptions. Companies like Cisco have been very successful in turning the education of their products into successful business. But these are exceptions. Even the exceptions have one goal in mind—to support company profitability. If your company produces and sells a product, service, or solution, the success of your company is measured in the sales of those products, services, or solutions. What determines the success of your company should determine the success of your training department.

The first thing anyone considering developing a training program should ask is what company goal it will help to achieve. Never let your product training become an island when it comes to the goals of the department. Your reason for offering the training should support your company's main objectives, or it will be very easy for senior management to cut funding for your department or eliminate it all together. This is true whether the training department is run as a cost of doing business or as a profit center. All costs of doing business must have a return to justify the expense. The fact that your department is covering its costs in the short term is of little consequence. Remember, most for-profit training centers are making what amounts to a very, very small percentage of the overall profit for their company. Their real task is to justify their existence.

Technical training departments that focus on the bottom line will be less focused on things like the modality of the training. A good curriculum designer will start with what they want the student to be able to do when they have completed the training and determine the best method to make that happen. It may be online learning, instructor-led, or a combination of both. Regardless, they will worry less about selling the training to make a profit and more about making sure that their training is increasing associated product revenues by more than the cost of doing the training. *That* is true profit.

It is interesting to me how willing executives can be to spend money developing learning when they have been promised that those costs will be recouped through the sale of the training itself, but are so reluctant to spend money when the increased sale of the product will recover the costs, sometimes many times over. The reason? Because getting accurate numbers for the first scenario is easy. Getting accurate numbers for the second scenario is difficult, and executives live by numbers they can prove.

Conclusion

Product education that increases the bottom line is never *just* product training. Trainers are salespeople. Just because the training may happen after the sale does not diminish the influence that instructors have on future sales. Even if the training content is heavily technical and the audience is not comprised of decision-makers, effective product training will influence sales and decrease support costs. If a company is willing to invest in training their people on your products, there is a good chance they will buy again. Your customers will always want more of what they know best. Your job, then, is to make sure your customers can do more with your product than they can do with your competitors' products.

Product education existing as a cost of doing business will only thrive when business is good and profits are high. Training that turns a small profit but does not increase the mission and value of the company as a whole will remain a training department only until the stakeholders demand increased profitability or a narrowed focus. However, successful training is training that makes your company successful. It will last even when times are difficult, if you can prove that through accurate and meaningful numbers.

Make sure your training objectives support your company's business goals. Link every class you offer to a specific business goal. Do not let the goal of teaching about your product be to cover your expenses. Even worse, do not let training become a post-sales necessary evil. If your courses and modules make your CEO successful, your training department is here to stay.

Making It Practical

For good training to exist in your company, it must be part of a larger strategy. Even if you are not involved in defining that strategy, it is crucial to know why your company wants your training to be effective.

1) How does (or will) the next class you teach help your company increase revenue, avoid costs, or improve services?

2) Describe, in your own words, how product training can become more than *just* training about your products.

3) How does training that uses the "profit center" model tend to minimize evaluation of behavior change?

Before you read Chapter 6, "From Good to Great: Defining the Focus of Effective Product Training," answer these two questions.

1) In your own words, describe the difference between facilitating and lecturing.

2) What is more valuable to your training program, individual results or number of students training?

Note

1 While the numbers vary between the resources, they all agree that many billions of dollars are invested in corporate training every year. In a recent study by the research firm Bersin by Deloitte, they found that US companies alone spent over 70 billion dollars in corporate training. The Association for Talent Development had a number much larger, quoting in 2012 that US companies spent 162 billion dollars in employee education. Either way, the numbers are huge! For more information, visit www.home.bersin.com and www.atd.org.

6

From Good to Great

Defining the Focus of Effective Product Training

Everybody wants to provide great training. I know you do, or you wouldn't be reading this book. You instinctively know that bad training is worse than no training and that you are better off not training your customers than to provide them training that is either ineffective or incorrect. The issue is not about separating bad training from good training. Usually, that is not too difficult to determine. What is harder for most people to define are the nuances that differentiate good (or adequate) training from great training.

The first few chapters covered learning strategies and philosophies that will help create a productive learning environment. Those foundational principles are important to get right. The details are also important to get right. Having good educational and business philosophies is important, but can be wasted if the details are not addressed. This chapter will address nine contrasting areas that you should consider when progressing your training from good to great. There is no particular order to these nine items nor are they all equal. Your specific circumstances may require you to come up with your own contrasts, as they apply to you and your industry.

Aim at the Right Target: Doing Versus Knowing

Remember: You are valuable for what you do, not what you know. The smartest person in your company can be let go tomorrow if they either do something stupid or fail to do anything of value. This is not a new concept, and you are likely to find trendy statements all over social media that state similar ideas. What should interest you is the application to training and, more specifically, product training.

If doing is what makes you valuable, then doing is what makes your students valuable. Too often, instead of concentrating on doing, smart people put the emphasis on knowing. Instead of teaching to increase proficiency, most technical experts teach to increase knowledge. When knowledge is your end goal, the best you can do is hope some of what you teach sticks in the minds of your students. When doing is your goal, you can teach with confidence that you can achieve your goal. Make sure you aim at the right target. Know what you want your students to do differently and find a way to make that happen.

Product Training for the Technical Expert: The Art of Developing and Delivering Hands-On Learning,
First Edition. Daniel W. Bixby.
© 2018 John Wiley & Sons Ltd. Published 2018 by John Wiley & Sons Ltd.
Companion website: www.wiley.com/go/Bixby

Change the Approach: Facilitator Versus Lecturer

If it is true that you cannot transfer your knowledge to your students—that they must learn the knowledge for themselves—then it is also true that your job is not just to deliver a lecture. Effective instructors will significantly reduce lecture time, though they won't abandon it completely. The important thing is to enable the student to make the material their own. The key is to facilitate learning. The word "facilitate" comes from the French verb *faciliter*, which means "to make easy" or "to simplify." That is the job of proficiency trainers. You must make the learning as easy as possible for your students.

A good facilitator enjoys the challenge of taking the most complicated technical product and making it easy—for the student. An expert revels in making it look easy—for themselves. An expert accentuates the gap between their knowledge and their student's knowledge. A facilitator bridges that gap, often without the students ever realizing it. Students don't forget experts, but rarely remember what the expert taught them; they often forget a facilitator, but remember both the learning environment and the skill they were taught. The facilitator doesn't care if they are remembered. They are more concerned about their students than themselves.

Facilitators are concerned with the learning environment. They want their students to be comfortable. They want their students to have fun. They understand that content is important, but it is not king. They know that icebreakers and games can make understanding easy and that is their goal. Experts, on the other hand, want to skip "the fluff" and get to the "real reason" the students came. They fail to realize that when they do that, they are merely delivering content, not learning.

Facilitators are concerned with all of their students, while experts focus on only the ones they perceive to be more advanced. A facilitator is deeply grateful that their students have taken the time to come learn from them. An expert expects their students to be grateful for the time he or she is giving to teach them.

The task of making learning easy is not easy. The challenge, however, is extremely rewarding. It is a skill that must be practiced to improve. Anyone can be given the task of teaching. The real assignment is to be an expert teacher.

As you've probably noticed, the issue is really one of attitude. If you are reading this book, you are on the right track. A facilitator wants to learn how to be a better instructor, while an expert is, well, they're already an expert. As a subject matter specialist, you are at an advantage. If you can learn to be a facilitator, you can be a great instructor. Subject matter expertise is important. The more you have, the further you can take your students. But it is only helpful if you can also become a great facilitator.

I can often take a great facilitator that is new to a technology and turn them into a great technical instructor faster than I can take a subject matter expert and turn them into a great facilitator. I don't state that to discourage experts from becoming trainers, but simply to emphasize how difficult it is to let go of what we feel defines us–the knowledge gap between us and them. But since value is not established by what you know, but by what you do, and your job is to do training on your product, you become very valuable when you do training well.

Call It the Right Thing: Training Versus Presentation

What you call the educational experience you are trying to give your audience matters. Most business and technical professionals deliver presentations, either formally or informally, on a regular basis. It may be in a small, internal meeting environment, or in front of hundreds of customers, but most will deliver some type of dialog with the intention of changing or enhancing ideas about a product, service, or methodology. All too often, that exchange of ideas is wrongly called training.

A presentation is not training. Companies are willing to spend a lot of money teaching people how to present, but few teach their subject matter experts how to train. Learning to present is important for trainers, but it is only part of the necessary training instructors need. When a subject matter expert gives a presentation, they are merely conveying information to a group of people. If that is your goal, do it. But don't use the same material to teach a training class. If you do, you will be subjecting training to the same expectation as a presentation, namely, one of awareness.

When technical products are the subject matter, it can be even more difficult to make that distinction. Technical experts are notorious for using the word *training* to imply a depth of knowledge beyond a mere presentation. Apparently, when a presentation grows up, it becomes training. The distinction, however, should not be about the depth of the knowledge. I've heard tremendous and incredibly technical presentations. I've also seen excellent but very simple technical training. The terminology has nothing to do with the content. It has everything to do with the desired outcome.

Training is about doing. Presenting is about knowing. Too often, instead of concentrating on doing, instructors emphasize knowing. Instead of teaching to increase proficiency, many technical experts teach to increase awareness. Of course, increasing knowledge is important. Almost all training will have presentation elements built into it. But not all presentations have training built into them. The main difference is whether your objective is to change behavior or to enhance knowledge. If your goal is to drive proficiency, make sure you aim at the right target. Know what you want your students to do differently and find a way to make that happen.

Presentations are an effective way to motivate and encourage one to take action or do something. However, if that action involves a skill, training may be required to verify that the individual can actually perform the skill.

There are other ways to help determine the difference between a presentation and a training session. For example, a presentation is usually a better way to motivate and encourage change, while a training class requires demonstrating that change.

A presentation can be effectively delivered to hundreds or thousands of people. Or none. The audience size is irrelevant in determining whether you delivered a presentation well. Many excellent presentations have been delivered in front of a camera with no audience at all. Some went on to be viewed by millions; others have never been seen. How many people view that the delivery influences the reach but has little to no influence on whether it was delivered well or not? While the size of the audience will certainly influence certain aspects of a presenter's delivery—such as their voice projection, where they look, certain gestures, and body movements—it will rarely alter the main objective of the presentation. The objective of a training class, however, does change along with the size of the class. Only the simplest of training objectives can be taught to large audiences. Technical training classes are best delivered to small groups of individuals. In order to allow more time to practice and demonstrate the objectives, fewer numbers are usually required. As a proficiency trainer, you must make decisions about how many students you can effectively train on your product.

A presentation can be validated almost immediately—often by the obvious acceptance of the audience through applause or other means. Immediate survey results can provide instant feedback on the success of a presentation. More often than not, a presentation is successful if people enjoyed it. Training can be very successful even when people don't like it. Of course, most instructors want people to like their training class as much as they want them to like their product. A person's reaction to the class, however, does not determine its success. Some students will love a class and give it high marks, but their performance will not change very much at all. Others may be critical in the beginning and eventually change their behaviors as a result of the class. The point is that training takes time to validate.

Presentations focus on the delivery. The word itself implies that the emphasis is on how the material is *presented*. A well-delivered presentation is a powerful tool that can inspire good endeavors or dissuade bad actions, but it is ineffective when teaching a skill. Training, on the other hand, focuses on the receiving. What matters most in training is how the learning is internalized by the student. Verbal communication is still important, but only insomuch as it helps to change the way a student does something.

While presentations build on the presenter's experience, training builds on the students' experiences. This is a key difference. In order for learning to take place, it must build on the learner's experience. You have likely forgotten almost everything you've heard at the hundreds of presentations you have attended in your career. I wish I could say that your training classes have been different, but so many of them were likely presentations that were called training that there may be little difference. If you have attended training that encouraged you to construct the learning for yourself through hands-on activities and exercises, you likely retained a significantly larger portion of that training.

A training event is also different from a presentation event. While learning can certainly happen at a presentation, it often *is* the learning event. A good presenter can

Table 6.1 Presentation versus training chart.

Presentation	Training
• Successful when it changes or enhances knowledge—about knowing	• Successful when it changes or enhances a skill—about doing
• Motivates and encourages change	• Demonstrates change
• Audience size is irrelevant	• Learning is individualized
• Validation can be immediate	• Validation may be lengthy
• Focus is on the delivery; how it is *presented* and builds on the presenter's experience	• Focus is on the receiving; how the learning is internalized and builds on the student's experience
• Often, it *is* the learning event	• Generally, is one step in a process

introduce, deliver, and wrap up the learning—all in one inspiring presentation. A training event, however, is usually one step in a process. The process can be short or very lengthy. Training can be condensed into one event comprised of multiple parts, but it is almost never a single presentation.

The chart in Table 6.1 may help to condense the aforementioned distinctions between a presentation and training.

Make It Sustainable: Standardized Versus Customized

Differentiating between a presentation and a training course will help with consistency and sustainability. Both are powerful tools when used at the right time. The same is true of standardized training and customized training. Both are good, but only when used appropriately.

If you are a subject matter expert, you likely have multiple PowerPoint files on your computer about your product. You are ready to present at a moment's notice. If a customer calls up, you tweak your presentation, delete the slides you think they don't need, and add a few others, and you're ready to go. Your coworkers all do the same. Soon, there are many versions of the "training" that you and your colleagues are using to educate your customers on your product.

Customizing training is a great thing to do. But sustainable training must remain consistent to its objectives. Customized training happens when the course objectives are designed for a specific customer. You cannot, however, simply change a few slides to match what you want to tell a customer and call it customized training. That may well be a personalized presentation—and it may be powerful and effective. But you must differentiate between a personalized presentation and a sustainable training course on your product.

The problem with subject matter experts having multiple, independent training courses is not in the individualization of the training. As noted earlier, individuals learn, not companies. Training must be personalized or it won't be effective. But using an individual's past experience to help them learn a consistent objective is different from changing the objective of the course for every class. The problem is in the inconsistency—or more precisely, in the unintentional inconsistency. If you are going to offer customized training, you must deliberately and precisely customize the objectives and the activities as well, not just the content. Make sure you have the system and means to keep track of the various versions and what the differences are.

Measure the Right Things: Performance Versus Reactions

It's one thing to aim at the right target and to expect good results; it's another thing to measure them, which you must do. You must be able to prove that you got the results you set out to get. You prove results by measuring what your students can do and if they can remember how to do what you taught them. You can also prove results by measuring the impact your students believe the training will have on their job performance. Simply measuring student's reactions to the training will not prove results. Asking them if the snacks were good may help provide better coffee at the next training, but it will not improve your ability to increase their job performance.

Chapter 7 will cover more about how to evaluate training effectiveness. Measuring the right things is listed here because it is one of the important deliverables required to improve your training from good to great.

Value the Right Things: Results Versus Head Count

Far too many training programs count their success by the number of students that attended a class. Tracking the number of attendees is important, but numbers don't tell the whole story. When head count determines your success, expectations and goals are flawed and ultimately lead to the demise of the training department or program. If seat numbers control your destiny, you will likely fall into the cycle of ineffective training. First, you need to fill the seats, so you fill them with anyone who will take the class. Because many of those students are not the right students for the class, the effectiveness of the training will be compromised. If your goal is to increase the number of students taught, however, your only option is to get even more students. Now you are stuck in the cycle that has ended many training programs. See Figure 6.1.

Having a thousand students go through your class is great, but how many of those were from your target audience? If they weren't from your target audience, do you know what demographic they are from? This could help you determine whether you need to create a separate course for them. Your success rate will also be affected since those outside of your target audience may not meet the intended objectives. When your training is judged by results instead of just attendees, you will demonstrate the true value of training.

Use the Right Delivery Methods: Effectiveness Versus Availability

Another contrast that will define your strategy is how you deliver the training itself. Will the training be delivered online or in a classroom? Will it be delivered live or will the student be able to access it any time? These are just some of the questions instructional designers ask when choosing the medium. To simplify the discussion, there are really two deciding factors that you must determine.

Figure 6.1 Ineffective training cycle.

Instructor-led versus learner-led. Training can be either initiated or led by an instructor, or it can be initiated by the student. This distinction is not one made by, or bound by, technology. The only difference technology has made to instructor-led training is where your students are. Now, instead of only being able to teach a class to students a few feet from you, you can teach one to students on the other side of the globe. The only difference technology has made to learner-led training is that it has opened up so many more options for accessing information. While libraries and printed books are still popular, so are online videos, podcasts, and mobile learning—all ways that a learner can initiate technical learning for themselves.

Synchronous versus asynchronous. Training is either delivered in real time (live), or it is delivered on-demand or self-paced. Like the contrast earlier, this distinction is independent of technology. A teacher can deliver synchronous training to students in a room or across the ocean. Training can be delivered asynchronously on paper or on iPads.

Your training classes will all be defined by these two contrasts. They will either be synchronous and instructor-led, synchronous and learner-led, asynchronous and instructor-led, or asynchronous and learner-led. In Exercise 6.1, put the training in the appropriate box. The answers are provided in Figure 6.2.

Exercise 6.1 Delivery method exercise

• Classroom
• Correspondence classes
• Online cohort classes
  (most university distance
  education)
• Livestream
• Video
• Live webinar
• On-demand eLearning
  modules
• Recorded webinar

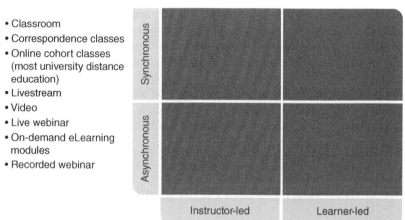

Figure 6.2 Delivery method exercise answers.

Blended learning. The term blended learning is often used to indicate that both classroom and online teaching will be used as a delivery method. Most often, that is what blended learning is, but it shouldn't be all about the technology. Blended learning is learning that incorporates more than one of the boxes earlier in the total learning experience. For example, a classroom instructor (instructor-led, synchronous) could have a distance education component (instructor-led, asynchronous), thus blending the two approaches.

Blended learning offers many advantages. When designed correctly, blended learning can offer the best of both worlds and enhance, not detract from, real learning. Offering self-initiated components to learning can be motivating to the student, especially if they are allowed to tailor the learning to their specific needs.

The important thing to remember is that the medium, or the way the learners get the information from us, should be a secondary decision, not a primary one. We should first determine the best way to deliver success—to accomplish the objectives we set out to do. Of all the contrasts in this chapter, this one likely has the most bias coming into the discussion. Many programs are built on a technology-first agenda. When those programs are backed by instructors and individual attention to the students, they can be very successful. Online university study has proven to be at least as effective as classroom study when the courses are properly designed. What makes the courses effective is that most are still instructor-led, allowing professors to make sure students understand concepts instead of merely memorizing facts. Unfortunately, however, most corporate programs are a one-size-fits-all approach. They worry about successfully delivering the training course to large numbers of people more than they do about effectively delivering success.

Proficiency instructors can use any medium to deliver on their objectives. They just understand that they may need to change the objective in order to do so. If the objective can't or shouldn't change, then they must have the flexibility to choose a delivery medium that will help them meet that objective.

Continue the Conversation: Process Versus Event

Training is more than an event. Proficiency trainers understand the importance of building on past experiences in the classroom. They also understand that the learning is not finished when the student leaves the classroom. Most students will learn a lot more outside of your classroom than they ever will in your class. That may be difficult to accept, as a subject matter expert instructor, but it is true.

Consider the progression in Figure 6.3. The challenge for instructors in a corporate scenario is to bring all of the experiences and training events together to promote proficiency. To do this, you must see training as a process. The event is just one step in the process. Some things you will have little control over and others will require creativity.

One way to continue the conversation of learning beyond the classroom is to use social media. You might, for example, create private LinkedIn groups for those who have completed a particular certification course. This could allow them to connect with their peers and continue learning from them. This works best when the conversation was started at the learning event. Making it easy for students to continue the connection after the formal learning is a practical way to continue the conversation and turn training into a continuous process.

Keep Improving: Progress Versus Contentment

Those who love learning are always improving. Good trainers make it their goal to learn something from every class and to do something better at every class. In a very real sense, your course will never be completed. As you learn from students what helps them most, you need to continue to make changes to your approach, to your exercises, and even to your curriculum.

I learn something at every class I teach. The challenge is to capture that learning and, if

Figure 6.3 Training event is one step in a process.

appropriate, use the new idea in future classes. Knowing when to require consistency and when to insist on change requires practice. The goal is to be continuously improving, which takes work. My motto is "Every class is better than the last one." Never let the fact that you've always done something in a particular way keep you from making an improvement. In most cases you'll be able to find at least one area that can be improved.

Unless you are completely revising the curriculum, don't try to change too much. Pick one or two things you want to do differently and try them. If they don't work, you don't have to do it again. It may be as simple as a new illustration or a new exercise. You may want to try teaching a particular concept by asking questions or having a group discussion. Whatever you do, document the change and make notes for yourself. If possible, have someone else there to observe it and give feedback. Don't be afraid to tell the students about the change and get their direct feedback as well.

Conclusion

All nine of these items take a lot of work and a lot of trial and error. Add in your own contrasts, and then prioritize them. Keep them as a goal, working on one or two at a time, as you seek to improve your training. Over time, you can move your training from good to great.

Making It Practical

Taking your training from good to great isn't difficult. You just need to focus on some of the little things. Don't let the small details become big roadblocks for great training.

1) What are you doing to ensure that every class is better than the last?

2) What makes you more valuable to your company: what you know or what you do? What makes your customers more valuable to your company, what they know about your product, or what they do with your product? How does that change the way you train them?

3) In your own words, describe the difference between training and presenting.

Before you read Chapter 7, "What is Expected Must Be Inspected: Assessing and Evaluating Hands-On Learning," answer these two questions.

1) In your own words, describe the difference between assessing and evaluating.

2) In your experience, how helpful have surveys been in improving future classes?

7

What Is Expected Must Be Inspected

Assessing and Evaluating Hands-On Learning

One summer, as a teenager, I worked with several other young people to help restore an old farmhouse into what would become a youth camp. Our supervisor, Mr Abuhl, was an older gentleman who had volunteered his time and talents and was intent on improving our renovation skills.

Many people were surprised at the quality of work Mr Abuhl was able to produce with a few 15–17-year-old teenagers. Those of us that worked for him were not surprised at all. His success came, not by cajoling more work out of us, but by constantly evaluating the work that we did. He would meticulously inspect our work to confirm that we were meeting his stringent (at least from a teenager's perspective) standards. "Work that is expected must be inspected," he would say. It was his way of telling us that his re-measurements and careful scrutiny of our work were not because he didn't trust us, but because detecting small deviations now could help prevent larger problems later on.

The same is true with training. Whenever you provide learning, you must verify that the training did what you intended it to do. If you don't, no one may notice for a while, but soon your training will be deemed ineffective. Even if it isn't and your training is amazingly effective, or generates tremendous revenue, few will care unless you can prove it.

In the 1950s, Donald Kirkpatrick suggested four areas that must be evaluated in any training program. They have become known as Kirkpatrick's evaluation model.[1] First, trainers should measure the student's reaction to the training. The traditional way many do that is through what we sometimes refer to as smile sheets or surveys done right after class to see if students liked the training. Second, you should determine if the individual students can meet the objectives of the course. The traditional way to conclude that is through a written test or an exam. Third, he suggests that instructors must measure behavior, or the change in their ability to do something specific. The traditional way to do this is through follow-up and on-the-job analysis after the training. Finally, trainers should validate that the class itself is getting the business results you want from the class. The traditional way to that is by demonstrating that you brought in more money than you spent.

Because proficiency training must include a change in behavior, or no real learning has occurred, I will combine levels 2 and 3. You must evaluate the course (level 1), you

Product Training for the Technical Expert: The Art of Developing and Delivering Hands-On Learning,
First Edition. Daniel W. Bixby.
© 2018 John Wiley & Sons Ltd. Published 2018 by John Wiley & Sons Ltd.
Companion website: www.wiley.com/go/Bixby

must assess the individual (levels 2 and 3), and you must measure the value obtained (level 4). All three are important to validate success.

It is a mistake to assume the Four-Level Training Evaluation Model is a sequential evaluation model. There are four different things being measured, but they don't always happen in that order. As I stated earlier, you must start with the end goal in mind. Also, you may assess the individual during a class and take a survey after the class. You may know what the business value is even before the students have time to demonstrate a change in behavior.

Earlier chapters dealt with evaluating the business goal of the class. This chapter deals with evaluating the class and assessing the individuals.

Assessing the Individual

In product training, there are two main types of individual assessments that you should consider: you should assess the student's overall knowledge and you should assess their skill level. Both are important in order to determine if you have increased your students' proficiency on your product. This is especially true if there is going to be any kind of proficiency certification granted.

Assessing Their Knowledge

Quizzes and exams are the most common way to assess for general knowledge. Both can be helpful to determine if students were able to process the material that you covered. Asking good test questions is a skill that may require seeking additional help. There are many resources available for learning how to write good quiz and exam questions. Some tools, like Questionmark.com, offer whitepapers and other helpful tools to create effective exam questions. This chapter is not a comprehensive guide to creating a quality quiz or exam.

Quizzes

Quizzes should be short and generally follow the module or lesson immediately, often using the same wording used in the lesson. The goal of a quiz is to make sure that students are keeping up with the information flow. Quizzes help you, the instructor, know if you need to cover something again, if you need to slow down, or if your students are with you or not. Quizzes can serve multiple purposes, like being a means of student engagement or preparing students for a later exam. Quizzes can even be given merely to take attendance. Whatever the reason, you should know what it is.

When you ask a quiz question, you should always review the answers with the class. Make sure that all of the students know the correct answer before moving on. I rarely grade a quiz, though students may see a similar question in the exam.

One way to include regular feedback, or quizzing of the students, is by using an electronic polling tool. A good polling tool or mobile application allows for all students to answer questions. Using these types of engaging tools will increase your effectiveness by adding interaction with all of your students. Another benefit is that it will force you to slow down and ask more questions.

Exams

Exams, or tests, are a more formal method of measuring a student's comprehension. Remember that some students are better test takers than others. Also, if you don't carefully evaluate your questions or requirements, there may be little to no correlation between a student's exam score and their skill level. In some cases, it may be necessary to forego a written exam altogether and concentrate on a competence or skill exam.

The best way is to use both. It is very likely that your driver's license required both a written exam and a practical exam. There are things like road signs and traffic laws that are important for you to demonstrate knowledge of, even if you are an excellent driver. The same is often true of your products.

If you do choose to create a written exam, here are a few tips that can help.

About Creating Exam Questions

There are a number of different types of question formats you can use to ask questions, and all have their strengths and weaknesses. There are also a number of things you should do when writing exam questions. Here is a list of a few of those:

1) **Get lots of feedback.** Whichever type of question you ask, it is important to get as much feedback as possible prior to using the exam question to affirm comprehension. Ask feedback from both subject matter experts and non-experts alike. It is important to verify both the clarity of the questions and the clarity of the answers. Getting as much input as possible will help you do that.
2) **Keep records of your exams and the results.** If students consistently get certain questions wrong, it could indicate a weakness in the curriculum or a poorly worded question. Questions that are always answered correctly may not need to be asked at all. One way to confirm that is to keep track of the questions and answers over a period of time.
3) **Include questions from each of your main objectives.** Make sure they are relevant. It is discouraging to the student when instructors ask questions about trivial data that has no bearing on their proficiency of your product.
4) **Ask questions that verify understanding.** Too many tests merely determine which students are better at recalling information. If it's not important, don't ask it. No one says your exam has to have a certain number of questions.
5) **Use multiple types of questions.** While a multiple-choice exam may be the easiest to grade, it may not be the best way to verify understanding.
6) When you do use multiple-choice questions, however, **make all the distractors plausible**. Distractors are the wrong answers in a multiple-choice question. Consider this example in a safety exam:
   If the fire alarm goes off, what is the proper response?

   A) Calmly exit the building using the nearest exit and proceed to your designated assembly area.
   B) Ignore it; they're usually false anyway.
   C) Hide under the desk and scream.
   D) Call or text a friend to find out what you should do.

   Answers B, C, and D are the distractors. None of them are plausible. This question, at least with these answers, is not a good test of comprehension.

7) If your class covers multiple days and a significant amount of material, time the exam but make it open book. This will make it clear that knowing how and where to get the information is more important than being able to recall it from memory.

8) Avoid trick questions. Your goal is not to see who is smart enough to pass the test. You need to encourage students to study for the right reason—to become proficient in your product, not to prove that they can outwit you or other classmates. Trick questions send the wrong message to students, indicating that the goal is to pass the exam not to comprehend the material.

9) If possible, weight the questions and even the answers. I like to use a tool that allows me to assign different values to the questions. Not all questions are equal, and not all answers to questions are equally wrong. This provides flexibility in asking simple or very difficult questions or including pilot questions and not grading them at all.

10) Make the test part of the learning. While an exam may come at the end of a learning event, you should not presume that the learning has ended. The exam is an extension of the learning. If students struggle with a question or concept, use the exam to help them. Give students reports on their exams so that they can improve in the areas they need to, even if they scored high enough to pass the class.

About Administrating the Exam

If possible, use a tool that allows for flexibility when writing your exams. Don't be afraid to use an online tool just because your class is offered in a traditional setting. Online exams are not exclusive to eLearning. They can just as easily be administered as part of a classroom learning event.

A good testing tool will allow you to ask multiple types of questions. Another important feature is question pooling. Pooling means that you create multiple questions for each objective and then ask for a random selection from each of the "pools." For example, if you have five objectives, you might create ten questions for each objective. Those questions are stored in separate folders. When you create the exam, you may require two random questions from each folder, or pool, for a total of 10 questions. This means that no two students will get exactly the same exam, which helps to eliminate cheating.

Don't feel like you have to hand out a certificate at the end of every class. This works fine when the certificate merely awards attendance, but it rarely works in a certification setting or when not all students are guaranteed to pass the course. Many online programs allow students to print their certificates, which may be the answer. Another answer is to simply mail certificates after you have had a chance to review exam and lab scores. While this creates an extra administrative step, it allows you another opportunity to connect with your students.

Assessing Their Skills

If you are not going to administrate an exam, you may choose to use one or more of the activities you designed in step 5 of the 4 × 8 Proficiency Design Model (see Chapters 8 and 9) as a skill assessment. These types of assessments can add significant value to your training, but they require significantly more effort to administrate. If you use a skill assessment, it is important to carefully define the skill you are assessing. Otherwise, you will be left guessing or hoping that students actually did the required work.

Table 7.1 Rubric.

	Points		
Skill	1–3	4–6	7–10
Software upgrade	Upgraded from v.2.3 to 3.0 with significant help or in greater than 10 minutes	Upgraded from v.2.3 to 3.0 with moderate help or between 5 and 10 minutes	Upgraded from v.2.3 to 3.0 with no help in under 5 minutes

Another item that is required is what educators refer to as a grading rubric. You don't need to call it anything fancy, but you do need one. A rubric is simply a measuring guide—often in the form of a spreadsheet—that ensures consistency and impartiality in your grading. A good measuring spreadsheet will still allow for some flexibility without being too rigid. The example in Table 7.1 shows a very simple spreadsheet, where the skill being assessed is a student's ability to upgrade a software program.

Notice both the flexibility and the detail—the combination of both subjectivity and objectivity. This ensures against the extremes but allows for instructors to weigh in on the nonmeasurable items as well. "Significant" and "moderate" are both subjective measurements. What the students are doing (upgrading from version 2.3 to 3.0) is very specific. The timing has elements of both, giving you a little flexibility without having a new column for each possible minute.

A grading matrix is not only a helpful tool; it may be a requirement if you decide to create a certification course. Even if an instructional designer is creating the grading matrix, you will likely need to assist them. You are the subject matter expert, and you know what should be measured and what shouldn't. It may, however, be easy to get caught up in too many details. Keep your grading as simple and practical as possible. A tool that is too complicated to use never gets used. The end result should be a consistent determination of who is qualified to do the job and who needs more practice.

Creative Assessments

As long as you can stay consistent in your measuring of students, you can get creative in your assessments. A practical exam may simply consist of a series of troubleshooting problems that students had to complete. It may include a check-off list of items for students to self-grade their capabilities in certain areas. In many product training settings, these types of assessments are fine to use, as long as you don't treat some students more favorably than others.

There are other ways, of course, to test your students. Exams don't have to be a series of written questions. They can be delivered orally, or it can be written as a paper or thesis. Remember, however, that this is product training, not a graduate degree!

Combining the Grades

Whenever possible, don't use only the knowledge assessment as the pass/fail indicator for the course. If only one can be used, it should be the competency exam, but generally the grades should be combined. There is not a set formula or weight that is used when combining a written exam and a practical exam. It all depends on your product. This is one of the reasons why you, the expert, are involved in the training on your product. What is important, however, is consistency. You must be able to apply the same rules to everyone.

Take the time to write out the process, and make sure it will work for your most experienced trainees as well as students new to your product. If you decide to allow past experience to qualify in place of a practical assessment, be sure to qualify the expectations. Keep the process as simple as possible, but detailed enough to maintain a fair standard.

Evaluating the Class

Evaluating whether students pass a class is easy enough. Evaluating whether the class is really teaching the right thing and helping students to improve their performance and proficiency is quite another.

If you are like I was for a many years, you may administer surveys because you're expected to, not because you see any true value in them. The results can be confusing and difficult to discern between what is good and what is bad (is a 4.2 average bad and a 4.4 average good?). You may be a better judge of whether a class went well or not than your students. When this is true, you will be tempted to throw out results or even stop doing surveys altogether. There is no need to waste your students' time and force them to quickly fill out one more thing before they can leave.

But if you truly want to make every class better than the last one, you must know what needs to be improved. Continuous improvement doesn't happen without a structure and plan. Eventually, you will need some evidence of improvement, or at least some ideas for areas to improve on.

The problem lies in the typical Likert-like survey question, as shown in Table 7.2. In the training industry, this type of survey is often referred to as a "smile sheet." It merely asks the students if they liked certain things about the class or not.

Kendall Kerekes and John Mattox of CEB[2] suggest ditching the useless "smile sheets" that merely ask if a student liked the training. They suggest using "SmartSheets"—evaluations that predict performance improvement—instead. They offer these alternative questions:

- Did you learn new knowledge and skills?
- Will you be able to apply them?
- Will your performance improve due to training?
- Will you have managerial support?
- Will your improvement improve the performance of the organization?

Table 7.2 Likert-like survey.

On a scale of 1–5 (5 being best), please rate the following:					
Content of the class	1	2	3	4	5
Length of the class	1	2	3	4	5
Instructor's presentation skills	1	2	3	4	5
Others	1	2	3	4	5

Those are big improvements. They begin to move out of the reactionary and toward application. In his book *Performance-Focused Smile Sheets: A Radical Rethinking of a Dangerous Art Form*,[3] Dr Will Thalheimer suggests, as the title indicates, an even more radical approach to gathering feedback. If you are involved in writing survey questions for your training, I strongly recommend the ideas he promotes, all backed by research.

Dr Thalheimer highlights several issues with traditional smile sheets and with the Likert-like scales they usually use.

Regarding smile sheets:

- **They are based on subjective inputs.**
  Problem: While subjective input can be helpful, it can be inaccurate. We should, according to Thalheimer, be skeptical of the findings. (I guess I was right to be skeptical!) What one person feels they "agree" with, another will mark as "strongly agree," even though they both feel the same way.
  Solution: Instead of using subjective answers, give the student objective options, or at least more descriptive answers, as in Table 7.3.
- **They are usually given immediately after training and in the context of the training environment.**
  Problem: Learners are more biased toward training that is still fresh in their minds. They will also better recall the learning because they are still in the same context they learned it in.
  Solution: Survey students more than once. Ask them immediately so as to capture information that you might otherwise miss, and ask them questions after a delay so as to eliminate the bias created by the recentness of the training.
- **They often ask the wrong questions.**
  Problem: If this is true in environments run by professional trainers, it is even more true in environments where subject matter experts are leading the training.
  Solution: This is the problem that Kerekes and Mattox were dealing with, and one answer is to change the questions, as they have suggested.

Table 7.3 Evaluation question (question developed by Will Thalheimer, PhD, of Work Learning Research, Inc., based on his book *Performance-Focused Smile Sheets: A Radical Rethinking of a Dangerous Art Form*. SmileSheets.com.).

Now that you've completed the course, how able are you to TROUBLESHOOT the MOST COMMON ISSUES related to the H-Inx Distributed Antennae System (HIDAS)?

A) I am NOT YET SKILLED in troubleshooting the most common HIDAS issues.

B) I have GENERAL AWARENESS of HIDAS troubleshooting, but I WILL NEED ADDITIONAL GUIDANCE to ensure a complete and reliable fix.

C) I CAN HANDLE A LARGE PERCENTAGE of the most common HIDAS troubleshooting situations, but I will need ADDITIONAL GUIDANCE to ensure I can fix all of the most common issues.

D) I CAN HANDLE ALL of the most common HIDAS troubleshooting situations, but I still need ADDITIONAL GUIDANCE on LESS COMMON issues.

E) I CAN HANDLE ALL HIDAS troubleshooting situations, including the most common and the least common issues.

Regarding Likert-like measurement scales:

- **They create poor decision making.** Thalheimer argues that when students answer survey questions, they are making decisions. Having to choose between *very satisfied* and *extremely satisfied* is a poor decision.
- **They create ambiguous results.** I can attest to that! I've seen bad instructors and ineffective training get good results and great training get questionable results. Give a bad training class in Hawaii and you are probably guaranteed to get "highly recommended" results!

Table 7.3 offers a sample question, reprinted here with permission from Dr Thalheimer.

Evaluating Perceptions

To be clear, measuring reactions is not always bad. In fact, from a marketing perspective, it is a good thing to do. If you are training customers, for example, you want to make sure that they enjoyed the time that they committed to you and that they will return or send others back. Maybe the coffee you serve is that important to them. But measuring reactions captures the student's immediate impression. You must do more than just measure immediate impressions. You must measure the change you've been able to make in their behavior and/or performance.

Ask questions that force the student to move beyond the reactionary to the use. Even if the questions themselves are not as objective as an exam question, the data is still helpful to predict the effectiveness of the training, if you ask the right questions.

Here are a few additional tips about survey questions. These come from my own experience and are subject to change, depending on your circumstances. For professional help creating questions that will drive performance change, you should engage a company like CEB or Work Learning Research that specializes in gathering the right data.

Ask as few questions as possible. Don't ask fewer than you need to, but don't ask any more than you must. Too many questions will make the students rush through the answers.

Ask everyone. While fewer questions are a good thing, fewer people are not. If you want to start building meaningful data that can drive more effective training, ask everyone that participates in your training to fill out a survey.

Don't ask questions about something you won't or can't change. If you have no intention of changing the temperature in the room, don't ask the students about it. You are wasting their time and yours and diluting the survey.

Have a purpose or goal for every question. This is similar to the statement above, but has a slightly different application. You may be asking a question to gather data or to set up a benchmark for further training. You may want to verify that your training class aligns with company goals or with your philosophy of education. Whatever the reason, make sure you have one.

Don't ask leading questions. It won't take too many classes before you'll realize that you can manipulate answers. The goal is real data. Data that makes the program, instructor, company, or product look good at the expense of improvement is not helpful.

Allow for anonymity. Depending on the questions you are answering, it may be important to allow students to remain anonymous. This is especially true if you are asking questions about the instructor's facilitation skills, as mentioned below.

Ask to quote them by name. This may seem like a complete contradiction to the statement above, but it is not. Ask the students if they would like to make a statement about the class. Make sure you allow for the statement to be recorded separately from their anonymous survey. Statements from students can lend credibility to your class and you can often learn from them.

Provide an opportunity for written responses. Written responses are much more valuable than a checkmark or circled response. Ask them to tell you the one thing they would do differently in the next class, but also ask them what they would do the same in the next class. Let students know that you don't want to stop doing something they think is helpful. Get feedback on what you are doing right, not just what you are doing wrong.

A Note about Measuring Instructor's Facilitation Skills

Surveys are not exams. They are not exams for the student and they are not exams for the instructor. Too many times students get the impression that their survey is the instructor's assessment. If the instructor has done his/her job, he/she has begun to build a professional relationship with the students. As a result, the students want to help the instructor. Now you have skewed data.

Ask questions about the instructor's facilitation skills under two conditions:

1) The goal must be larger than getting a particular "score" in one class. You may be changing the style of teaching and want to know how students adapt to it. You may be wanting to create an environment of continuous improvement over a long period of time. Those are good reasons to evaluate the instructor's ability.

2) The survey must be delivered in such a way that the students feel comfortable that the instructor will not see their answers. When an online survey is not possible, ask an administrator or someone other than the instructor to give out the survey. Let them collect the surveys and put them in an envelope while the instructor is not present. The goal is to make the students as comfortable as possible writing their true opinions.

Conclusion

Surveys and course evaluations are here to stay and I have a renewed faith in their ability to help improve the effectiveness of product training when administered correctly. To be effective as an instructor, you must measure results. You must measure that your students can perform the objectives you set for the class, and you must measure that the class itself was effective in making that happen.

Making It Practical

The reason you take surveys and give exams is to make sure that the training you are offering is effective. Good surveys can also ensure that you are always improving your training courses and listening to your students.

1) In your own words, describe the difference between a Likert-like question and a performance-focused survey question.

2) Describe the difference between assessing an individual's knowledge and assessing their skills.

3) In your own words, and applied to your own setting, list the benefits of using a survey to evaluate a class.

Review of Part Two: The Strategy of Hands-On Learning

1) Which of the principles in the last three chapters might you use to improve the product proficiency strategy in your own program? Which one(s) confirms a process or strategy that you are doing well?

Notes

1 Kirkpatrick, Donald and Kirkpatrick, James D. Evaluating Training Programs: The Four Levels, Third Edition, San Francisco, CA: Berrett-Koehler, 2006.
2 CEB, Five Myths about Measuring Training's Impact, Arlington, VA: CEB, 2015, p. 9.
3 Thalheimer, Will. Performance-Focused Smile Sheets: A Radical Rethinking of a Dangerous Art Form. Somerville, MA: Work-Learning Press, 2016.

Part III

The Structure of Hands-On Learning

8

Dethroning King Content

A Paradigm Shift

As an expert in your company's solutions, one thing you are good at is providing content. You may believe in the importance of good adult education principles and effective delivery methods, but still feel that they are merely colorful additives to what really matters—the content. They are not. Yes, it is true that when any of those things is missing (including content), there is sure to be a void in the quality of your product solution training. And yes, it is also true that the easiest way to fill that void is with more of what you know best—content.

In most technical training courses I have taken, content is the undisputed king of training. I'm not sure if it was by a coup d'état or a democratic election, but somewhere in history content was promoted from the ranks of informant to that of supreme ruler. Its entourage of defenders and promoters are many, and your battle won't be easy, but before you can move from a good training program to a great one, you must first dethrone King Content.

Now, don't get me wrong. Content is still important. This isn't some kind of French Revolution. I'm not calling for Content's prompt execution. Content has done nothing wrong and still deserves a very prominent place in your training kingdom. In many monarchical countries, the king or queen is a figurehead that is less powerful than others in their government with lesser titles. Similarly, content becomes even more powerful when prioritized correctly. A great training course will be less about what we teach the students and more about what the students can do when they are finished. Content serves best when it serves learning.

All trivial personifications aside, my guess is that content, or the material you provide your students, rules your training program. The reason content is so important to you is because it's the one thing you know you can get right. You own it, it is yours, and you want to display it. The content makes your training unique. Therefore, you reason that the content makes your training valuable.

Product Training for the Technical Expert: The Art of Developing and Delivering Hands-On Learning,
First Edition. Daniel W. Bixby.
© 2018 John Wiley & Sons Ltd. Published 2018 by John Wiley & Sons Ltd.
Companion website: www.wiley.com/go/Bixby

When Content Is King

When content is the only important emphasis of your curriculum, you will make poor learning design decisions. It is easy to appoint a subject matter expert to the task of teaching a technical subject without considering how learning is truly going to take place. It is important to understand that while material can be distributed, knowledge cannot. More PowerPoint slides may mean more material, but it does not mean more knowledge. There is a big difference between presenting facts and the comprehension of those facts. I've already emphasized that it is not enough to teach students to parrot facts. You must change behavior. Students can take a training module and immediately repeat back the answers to the questions you have chosen to ask them, but will they lead with your product six months from now? Will they remember how to use or install your product or troubleshoot potential problems? If the education you are providing them is going to last beyond the classroom, you must start with solid learning principles that maximize long-term retention and future development. Training that is merely a data dump of information will have little value beyond the training event itself.

The idea that content is king is a marketing one. In the early days of the Internet, Bill Gates wrote an article titled "Content Is King" in which he promoted the idea that news agencies with the most content would own the Internet. Of course, his was not a treatise on how humans learn, but rather a not-so-veiled promotion of the soon-to-launch MSNBC cable network and interactive online content. After a generation of Internet use, we should know that content alone is not king. Content must be accurate and relevant.

Advocates of the content is king concept agree. In fact, they argue their position on the basis that the substance you teach must be accurate. Learning, they say, is no good if the content is incorrect. Of course, they are correct in their assessment, but it does not mean that content is preeminent. That same logic can be said the other way around; the best material in the world is no good if your students don't learn it.

Many technical training departments have failed because they can't grasp this concept. As a technical specialist or instructor, your responsibility and profession is to make sure that learning happens. Content alone will not always accomplish that. Not always—but it might.

What if Content *Is* All They Need?

Perhaps there are exceptions to every rule. You might think that's what I'm doing—making an exception about how adults learn—by acknowledging that sometimes content is all that is needed for learning to take place. I am not. Learning still happens when we take new information and construct it on existing knowledge and experiences. But there are times when content is all you need to offer a student to make that happen. A facilitator is not always the key to learning.

Consider the dictionary. You don't need a facilitator to look up every word you want to learn about in the dictionary (online or on paper, it doesn't matter). The dictionary is a study help, a tool, or a job aid. Now, it likely took a facilitator to teach you how to use the dictionary, but now that you know how to use it, that facilitator—or elementary teacher in this case—can gladly back off and let you learn by accessing content alone.

The dictionary is a good example because it was designed to be used in just that way. You don't read a dictionary. You use the dictionary. You first provide the context and experience to learn and then you use it to find what you already know you need to learn. In the same way, content that is going to be consumed in that way must be designed to be consumed that way. This book does not address that, though there are many helps on building knowledgebase content.

How to Tell if Content Is King

So how do you tell the difference between knowledgebase content and content that has become the king of your training efforts? Diagnosing the problem is not difficult. Do you use the words "training" and "content" synonymously? Do you find yourself saying things like "I have the training, but I haven't taken it yet" or "I sent them the training" or similar statements? If so, you have probably assumed that handing over material is the same as the process of transferring knowledge. You may understand that they still have to consume the content, but even that would not constitute learning.

A dictionary is a powerful tool to a person who needs to speak, write, or listen. But a dictionary will never help anyone do any of those things. The most powerful thing about a dictionary is also the one thing that makes it a terrible curriculum. The dictionary is for everyone and includes everything.

In the same way, when content is king, your training will include too much information. When content is king, it does what it wants to do. And it wants to be seen by all. If you have a one-size-fits-all approach to training, it is likely that the material is more important than the message.

Another problem that arises when content is the most important part of your training programs is that you lose control of the content itself. If there are 20 product experts in your company and each of them has their own version of a "training" curriculum on their laptop, you have one of two things. First, you may not have training at all, but may merely have product presentations. Second, you may have training that is content driven. Content must be assigned its proper scale and context in order that you have more control over how that content is used and applied.

Giving Content Its Rightful Place

The most important reason to put content in its proper place is to get it right. I am not anti-content. I am so pro-content that I want to get it right. Compare your preparation to teach a product training class to your preparation to take a road trip. Of the questions listed below, which would you want to know first and which would you want to know last?

Exercise 8.1   Planning a trip exercise

What will you do when you get there?	Where are you going and who needs to go?
What do you pack?	How will you get there?

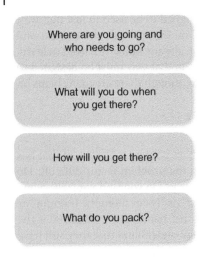

Figure 8.1 Curriculum design is like planning for a trip.

If you just found out you needed to take a trip, the first thing you would want to know is where you are going and who needs to go. Then you would probably want to know what you will do when you get there, followed by how you will get there. The last thing you will determine is what to pack (Figure 8.1).

In this analogy, what you pack is like providing the content. It is the deliverables, the things you will take with you on your trip (your training event). By putting it last, am I saying that what you pack is not important? No! In fact, take a closer look. Putting it last makes your packing more efficient and effective.

If you pack the car without knowing where you are going, you may pack skis and a snowsuit to take to the beach. If you don't know who is going with you, you may pack clothes for the wrong person. If you don't know what you will do when you get there, you're not going to pack the right equipment. If you don't know how you will get there, you may pack a trailer to bring on the airplane. The point is knowing that information in advance helps you to pack the right stuff.

The same is true with the content of your product training. More is not better. Packing all of the content in one class actually devalues at least some of the content, because you are stating that all content is equal, when you know it is not. If you didn't learn about your product in one training session, don't expect your students to learn it all in one training session.

Introducing the 4×8 Proficiency Design Model

When I build product curricula, I use an eight-step process (Figure 8.2). I have borrowed parts of this from several different sources, though I have modified it for product training specifically. I won't go into the detail of each step here. That will come later. I do want to introduce it, however, because this may be the single most important paradigm shift for you as a subject matter expert. You will get asked to provide training content. Make it valuable by putting it at the end of the curriculum design process. Content is so important it must come last.

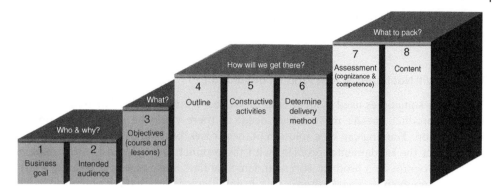

Figure 8.2 4×8 Proficiency Design Model.

The chart above shows the eight steps and four levels. The levels must be done sequentially, though some of the steps in the levels can be done out of sequence. The numbering shows the ideal situation, but flexibility in the levels leaves room for real-life situations. At the top of the table, you can see how it equates to the road trip illustration I used in Figure 8.1.

But before you spend too much time creating the perfect curriculum and designing the best learning possible, you must make sure that it will solve your problem.

Is Training the Solution?

When content is king, and *content* and *training* are used interchangeably, it may become difficult to determine if training will solve your problem. But training is not much different than the product you are training about. Your product or technology is successful when it helps to solve a real, a perceived, or a potential problem. Its credibility will be destroyed, however, if it is unable to fix the problem it was designed to solve.

The same is true for training. Training is not always the solution. One of the worst possible advertisements for training is training that does not work. And often training does not work because it is being promoted as the solution when it is not. One of the red flags that you may have content-based training is when training is offered as a solution for almost every problem. It makes sense. If you think material is synonymous with knowledge, and you're sure that a lack of knowledge is the problem, what more natural solution would you have than to give them more material? But more is not always better.

When you apply that to physical material, it is an easy concept to understand. Too much of a good thing can lead to disaster. Too much oil in the engine, too much air in the tires, too much paint on the brush, solder on the wire, time in the oven ... and the list could go on and on. When you apply that to content—the material you provide your students to learn—the same thing is true. It is impossible to have too much knowledge, but it is possible to have too much content. In fact, you'll see later how reducing content can actually lead to more proficiency.

For training to be successful, it must meet a particular need. The confusion comes in the word "particular." What is a particular need? It is a specific need for a specific audience. If you can't define the specific need and the specific audience, I would question the use of training as the solution.

Here are a couple of things to consider before you create a product solution training course.

Training Will Not Improve Your Product or Solution

Training is sometimes used as an attempt to correct a product flaw. The flaw could be that the product doesn't meet a true need, or there could be an actual technical imperfection. Training can be *a* solution in these cases, but don't ever assume training will correct the fundamental problem with the product. If you are asked to create a training program for a product with a fundamental flaw, make sure your training won't be judged on the outcome. In these cases, training programs are often perceived to be ineffective because—well, because they are—they've ineffectively trained customers to do something a product can't do.

There are times when training is essential as a temporary fix to a larger problem. I reluctantly accept these assignments as part of my job, but I admit, I don't like them. This is when training gets the "necessary evil" tag or executives get the idea that training won't be needed after the solution's flaw is corrected. Often, they are right.

From a customer perspective, your training gets the same reputation. It's never fun to teach a classroom or webinar full of people who know that they shouldn't have to be there. No matter how much time you are saving them, you're still wasting their time. If you have to train in this circumstance, make sure that there is a real solution in progress. If at all possible, do not offer a training course until that solution has been determined and progress is being made.

Training Is Not a Marketing Gimmick

I recently had the opportunity to visit Caterpillar's world headquarters in Peoria, IL. Sitting in the bed of a giant mining truck and watching videos of how it works was fascinating to me. But even though I learned a lot more about their mining trucks that day, I did not learn how to drive one. I am not, and likely never will be, proficient in using one. However, the trainer, or training video in this case, did meet its objective. It impressed me. Caterpillar never intended that I come away as a proficient operator of the CAT 797F. They recognized that the video was purely to provide me with information about the truck that would interest and entertain me. It worked.

Marketing and introductory videos are great ways to introduce concepts, but they are not sufficient in ensuring that a learner's behavior has changed. The strength of a video for marketing purposes becomes a weakness when it comes to training. You can use a video to reach thousands of people, with the hope of influencing a small percentage. You cannot accept those odds in a technical training class. Learning is an individual process, which means training is also an individual process. Effective training targets individuals. What is effective for training may not be efficient for marketing.

Creating a training course for the wrong reasons only dilutes the true value of training. You don't want your product to be used incorrectly and then be told it doesn't work. Similarly, you don't want your training program to be offered for the wrong reasons and then be told it isn't effective.

How Can You Know if Training Is the Solution?

So how do you know that training is the solution? Ask the right qualifying questions. Questions like *what* is the problem (or potential problem)? *Who* is completing this problem? *What* behavior/action needs to change? *How* will you know when it has changed? These and others are just some of the questions you should ask to determine if training is the right solution. Don't make up your mind that more knowledge will correct the problem without honestly looking for other solutions.

I recently received a request for training. After asking lots of questions, I realized the problem was not a lack of proficiency, but rather an issue with motivation. In this case, the problem was that technicians were cutting corners on an installation. What I found out, however, was that they were highly motivated to do so. They were being compensated by the quantity of installations without any concern for quality. In this case, a better solution would be to change the way the installers are compensated.

You cannot know if training is the solution if you do not know what is causing the problem or cannot articulate what success looks like. Make sure you can articulate those two things before you suggest that more education will correct the unknown.

I am not a great golfer. Actually, I'm not even a good golfer. But I'll never forget one of my better swings. Thankfully, I have a policy of only golfing with friends who swear to secrecy that I can hit the pitching wedge as far as I can the driver. On this particularly nice day I was feeling good. We were on the fourth hole of a very short par 3 course—exactly my kind of course. I must have only double-bogeyed the hole before and was feeling overly confident. I placed my ball on the tee and took a swing. It was perfect. The perfect club and the perfect swing. I couldn't believe it when my ball landed on the green, very near the hole. I had a chance at my very first birdie. My self-congratulatory yell of delight covered up the gasp from my friend. When I turned around, he was doubled over in laughter. I had almost gotten a hole in one—on the eighth hole. He still ribs me about that shot—though his version has gotten progressively better over the years.

The story illustrates that bad results start before the event (or golf swing, in this case) ever occur. Bad training starts with a misunderstanding of what training is and how adults learn. Good training cannot happen if you don't know what good training is. Even what might have been good, training is useless if it is not the right solution to begin with—if you're aiming toward the wrong green.

Figure 8.3 provides a flow chart by Cathy Moore that you may find particularly beneficial.[1] While her emphasis is not specifically technical or focused on product training, the applications are useful and helpful. If you follow it, you will avoid putting effort into the wrong solution. You will be able to answer with confidence if training really will help to solve your needs.

Conclusion

Putting content in its proper place does not minimize the importance of having good material. In fact, it turns good material into great material. It helps you know when to show a picture and when to pass out the actual product. It changes your learning event from a data transfer occasion into a knowledge building experience. Product proficiency instructors are never happy to simply transfer data. Getting content right means putting everything else first.

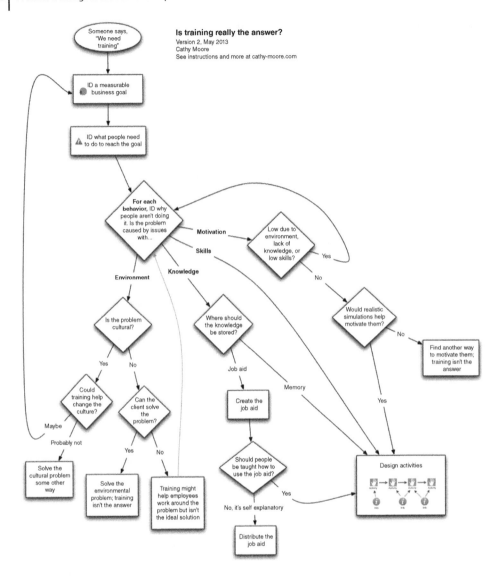

Figure 8.3 Is training really the answer?

This may be a paradigm shift for you. It has been for many engineers and technical experts around the world who have used the 4 × 8 Proficiency Design Model. It works. If you want to change what your students do with your product, you must have a process for dethroning king content.

Making It Practical

Adopting the concepts presented in this chapter may require a significant shift in approach for you. All trainers want their training to be effective, and it starts with getting the content right. In order to do that, it must be the last thing you do.

1) What two things should you be able to articulate before determining that training is the solution?

2) How does making content the last thing you consider when designing proficiency increase the value of the content?

3) Sometimes training is offered as a solution but does not correct the root problem. In your own observations, what are some of the possible effects of doing this?

Before you read Chapter 9, "Designing for Proficiency: Determining the Curriculum," answer these two questions.

1) What are the practical ways you design for proficiency? What processes do you use?

2) Describe the purpose of writing clear objectives before developing the content.

Note

1 Moore, Cathy. Is Training Really the Answer, 2013. www.cathymoore.com. Used by permission.

9

Designing for Proficiency

Determining the Curriculum

Think about the best training class you've ever delivered. More than likely, it didn't just happen the day the training was delivered. It started much earlier than that with a well-thought-out plan. In the learning industry, we call the process for creating that plan—or designing that curriculum—curriculum design. Curriculum design is yet another skill. It is a different skillset than facilitating effective learning, which is, as you are aware by now, a different skillset than product expertise. This book will not help you become an expert in curriculum design. However, there are three reasons why this chapter is very important for any good product facilitator.

First, as a product expert, you will get asked to provide content. You must be able to articulate how existing knowledge gets put into a teachable structure. If you can't, you will get stuck pulling existing information, adding material you believe to be relevant, and calling it training.

The last chapter covered what may be one of the biggest insights required to deliver effective learning. Namely, you need to make sure that content is not king. That is the second reason this chapter is so important: it provides practical solutions to help put content in its proper place.

The third important reason for this chapter is that it breaks down the four levels and eight steps of the 4×8 Proficiency Design Model and explains how they work together to create effective learning.

There are several curriculum design models that exist. They have been tested and used significantly. The most prominent one is the ADDIE model (Figure 9.1). ADDIE is an acronym that stands for analyze, design, develop, implement, and evaluate. The model was developed by Florida State University for the US Army in 1975. It is the most common instructional system design (ISD) model, with many variations to fit specific needs. As a high-level template, the model works for product training, since product training needs to go through each of those five progressions. This chapter deals with the two D's of ADDIE, designing and developing the training.

If you are working with a professional instructional designer, it is likely that she or he uses a model like ADDIE. I encourage you to use the model they suggest. However, when you get to the "Design and Develop" phases—or whatever their equivalent is called—you may find that you need more direction on pulling the right information together. As the product expert, you may feel like you are spinning your wheels, as if you are answering the same questions repeatedly. And you may be. The problem is in the

Product Training for the Technical Expert: The Art of Developing and Delivering Hands-On Learning,
First Edition. Daniel W. Bixby.
© 2018 John Wiley & Sons Ltd. Published 2018 by John Wiley & Sons Ltd.
Companion website: www.wiley.com/go/Bixby

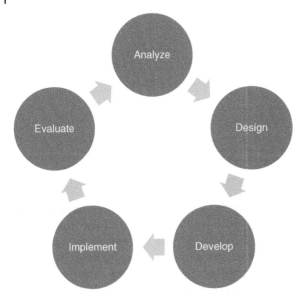

Figure 9.1 The ADDIE design model.

ambiguity between the design and development steps. Product experts need a simple template to help make sure the training content they offer on their product effectively increases proficiency.

The 4 × 8 Proficiency Design Model

The 4 × 8 Proficiency Design Model should be used with a comprehensive ISD. In some ways, it is more of a content preparation model. Analysis (or information gathering) is still important, as is the implementation and evaluation of your training. Evaluation, however, should not be treated as a single phase. Initial testing is a single phase, but continuous evaluation is not. Delivering a pilot class is a great way to thoroughly test a new training course or learning module. Evaluation, however, is ongoing. No two instructor-led classes should be exactly alike. The second one should always be better than the first. The three-hundredth should always be better than the two-hundred and ninety-ninth. Evaluation is not a phase that has a beginning and an end.

The 4 × 8 Proficiency Design Model has eight steps that are divided into four levels (Figure 9.2). The first level should stay the same for all lessons within the course, but levels 2 through 4 are iterative for each lesson. If your course only has one lesson, you should go through the four levels in order. If your course has more than one lesson, go through level 3 for the course (high-level objectives, high-level outline, high-level activities, etc.) and then go back to level 2. For each lesson, then, go through levels 2 through 4 until the course is complete.

Level 1

Business Goal

The first thing to do when designing product training is to articulate why you are providing the training in the first place. Sometimes, when I ask this question, I get

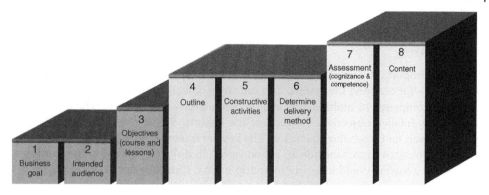

Figure 9.2 The 4×8 Proficiency Design Model.

either a very vague answer or a look that essentially says, "Uh … because they need it." I'm sure they do need it, but that is not a business reason for offering training on your product.

Knowing how your training will help your business is more than just an added benefit. It is mandatory. The urgency of a problem frequently clouds the importance of getting this step right. If you can't articulate how training will improve the services you offer, or how it will help you to either avoid costs or increase revenue, you are putting a lot of work into something that won't last long. You are likely well aware of what the business benefits are. You must take the tedious step to articulate them. Only then can you prove success. Guesses about what it might do or vague or hopeful statements about the benefits of more knowledge may make sense to you, but they are never going to convince executives about the need for training. Those leaders may allow a quick fix for an obvious emergency, but they are unlikely to let it continue as a long-term solution. Most executives are keenly aware that what you must do to fix an emergency is rarely the right long-term solution. Training programs that are born out of emergencies require the emergency to continue in order for the program to continue. In other words, there is not a real incentive to make the emergency go away. Always articulate and validate how the curriculum in question will benefit the company. This will prove the necessity of the class and help to articulate its value.

Intended Audience

While you are determining the business goal, you will also need to narrow down the intended audience. Who you want to train and why you want to train are often very much related. Don't document that your business goal is to sell more product but that your intended audience is internal field engineers. Be certain that the two align. It is, of course, possible—even advisable—to have a primary and a secondary audience, but make sure that the primary audience aligns with your business goal.

As you work to identify the intended audience, you may find that the audience affects your business goal. If it does, go back and change the goal until the two align. These must be right before you can move on to the next level. If, at any time, either the business goal or the audience changes, you will need to go back to the beginning and align them again. You must direct the objectives that you will work on next toward the target audience and for that business purpose.

Level 2

Objectives

The most important part of any training program is having a clear objective—or clear objectives—for teaching the course. You can write objectives at multiple levels. If your course is comprised of multiple lessons, you should have objectives for the course and then more detailed objectives for each of the lessons. Writing clear objectives takes practice, even for those who do it often. The key to remember with proficiency training is that the objective must include an action verb. "To do" is key. Avoid vague words, or words that a learner could interpret in multiple ways. Consider this vague example:

> "The successful student will learn about the different network cables."

As an objective, the statement is meaningless. One student could meet that objective almost immediately, perhaps as soon as they learn that more than one exists, while another may not easily meet that objective—if they are already bringing that knowledge into the class. "Learn more" and "Understand more" are two objectives technical experts often provide that you need to remove from your repertoire.

What you can do is replace that with a verb that demonstrates the level of knowledge you want to achieve.

> "The successful student will demonstrate their understanding of network cable types by choosing the appropriate cable for a given application."

Notice that in this example, it is clear when and how the student will meet the objective. The key is in the action verb, to demonstrate, which then requires another action verb stating how that demonstration will occur. The verbs must be observable (knowledge and understanding are not observable), and they must be measurable. What you will likely have to do is convert "topics to be covered" into actionable objectives.

Clear and consistent objectives will increase the flexibility of your training program. Customizing training is great. Changing objectives is not. Look at the examples above once more. Of the two examples, which allows you to be more flexible? Consider the first example again. If you arrive to a training site and find that all of the students are very knowledgeable about network cable types so you choose to skip that section, have you met the objective to provide learning about the different cable types? No. If you choose to cover it, but only quickly or peripherally, have you met the learning objective? Maybe, but not likely. If your students were already network-cabling experts, they likely didn't learn anything during that speedy session.

Now look at the second example. Even if all of your students are already experts, can you meet the objective and do so very quickly? Yes! You can have them demonstrate their understanding very early by choosing the correct cable types for given scenarios. The objective did not state that you would talk for 2 hours before allowing them to demonstrate their ability. In this case, the clear and observable objective made the class even more flexible and allowed for more time on other objectives or customized content.

Exercise

For the following exercise, consider three different business goals that a window manufacturer may have for training their resellers and observe how the different goals affect the audience and the objectives you choose.

A) Choose one of the following three business goals.

Exercise 9.1   Business goals

Reduce costs by cutting installation time by 20%	Increase sales by 5% by familiarizing resellers with the product and the rebate program	Reduce returns caused by faulty measuring

B) Now choose which of the following is likely the primary audience to meet that business goal.

Exercise 9.2   Primary audience

Reseller inside salespersons	Outside sales team	All installation technicians

C) Finally, determine which of the following objectives best fits with the pairing of business goal and audience you chose from above.

Exercise 9.3   Primary objective

Demonstrate proper installation of a 4-panel window in <90 minutes	Demonstrate order accuracy by correctly measuring and ordering windows for five given scenarios	Demonstrate an understanding of the rebate program by accurately choosing and filling out the forms.

If you don't know the business goal for doing the training in the first place, you won't be able to choose the right audience or determine the right objectives. Any of the following three scenarios are good reasons to offer training, but, when matched differently, shows the importance of making sure that all stakeholders are aligned in regards to the business goals and target audience.

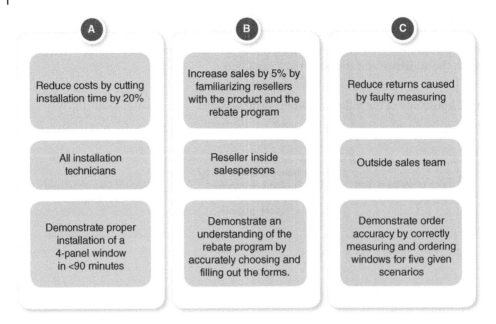

Figure 9.3 Answers to Exercises 9.1–9.3.

If you matched the three business cases with the appropriate audience and then chose the objective that best meets that business case, you likely came up with the following three scenarios. All are good reasons to provide training, but all three are very different and should be approached differently (Figure 9.3).

It is a mistake to go immediately to the objective and then work backward. I am very often given a vague request that goes something like this. "Reseller X needs to understand our product better. I told them we have a great training program and that you would provide them with all the knowledge they need." I'm grateful for the compliment, but it's only a great training program if we can tie each objective and audience back to the business goal. And my business goal is the same as the company's: sell more products.

Table 9.1 provides a list of verbs that is helpful to use when creating effective learning objectives.

Level 3

There are three steps in the third level. They can be done simultaneously or in a different order and may require some back and forth between them until you feel you have a good curriculum designed.

Outline

Making a good outline may require going back a few years into your days at the university and writing papers. Many of the same principles apply here. The key difference is that in a typical course, you have outlines within outlines. The first thing

Table 9.1 Action verbs for learning objectives.

A	Activate	Analyze	Appraise	Articulate	Assign
	Acquire	Answer	Approve	Assemble	Attain
	Adjust	Apply	Arrange	Assess	Attach
B	Balance	Bolt	Breakdown	Build	Bundle
C	Calculate	Check	Collect	Conduct	Coordinate
	Calibrate	Choose	Combine	Connect	Count
	Care for	Cite	Commission	Conserve	Create
	Carry out	Classify	Compare	Construct	Criticize
	Catalog	Clean	Compose	Contract	Critique
	Categorize	Code	Compute	Control	Cut
	Change	Collate	Conclude	Convert	
D	Debate	Describe	Diagnose	Discriminate	Distinguish
	Decrease	Design	Diagram	Discuss	Document
	Decide	Detect	Dictate	Dismantle	Draw
	Define	Determine	Differentiate	Dissect	Duplicate
	Demonstrate	Develop	Direct	Dispose of	
E	Edit	Establish	Examine	Explain	Express
	Employ	Estimate	Execute	Explore	Extrapolate
	Erase	Evaluate			
F	Fix	Follow	Format	Formulate	
G	Gather	Give	Grade	Green light	Grid
	Generalize	Glue			
H	Harvest	Highlight			
I	Identify	Improve Increase	Install	Interpret	Inventory
	Illustrate	Inspect	Instruct	Interview	Investigate
	Implement		Integrate	Introduce	
L	Label	Limit	List	Locate	Log
	Lift	Line	Load		
M	Make	Match	Mix	Monitor	Mount
	Manage	Measure			
N	Name				
O	Observe	Operate	Order	Organize	
P	Pack	Plan	Practice	Process	Propose
	Package	Plant	Predict	Produce	Prove
	Paint	Point	Prepare	Program	Provide
	Participate	Portion	Prescribe	Proofread	Prune
	Perform	Position	Press	Propagate	

(Continued)

Table 9.1 (Continued)

Q	Question				
R	Raise	Recommend	Refute	Replace	Restate
	Rank	Reconstruct	Regulate	Repeat	Restructure
	Rate	Record	Relate	Report	Retrieve
	Read	Recruit	Remove	Replace	Review
	Recall	Reduce	Renovate	Reproduce	Revise
	Recheck	Reflect	Reorganize	Research	Route
	Recognize	Refill	Repair	Resolve	Rewrite
S	Save	Select	Simulate	Sort	Structure
	Schedule	Separate	Sketch	Splice	Summarize
	Screw	Sequence	Skim	State	Support
	Score	Set up	Slice	Sterilize	Survey
	Search	Sharpen	Solder	Stratify	Systematize
	Secure	Simplify	Solve	Strip	
T	Tabulate	Test	Track	Translate	Trim
	Tape	Theorize	Train	Transplant	Troubleshoot
	Terminate	Trace	Transfer	Treat	
U	Update	Use	Utilize		
V	Value	Verbalize	Verify	Visualize	
W	Wash	Weld	Wrap	Write	

you must do is prioritize your objectives. It is perfectly fine for you to have more objectives than what you tell your students. For example, I teach a "train-the-trainer" class in which I list only five objectives, but I really have about eight. The three that I don't mention to the class are secondary to me. They allow me to be flexible and skip them if time or other circumstances don't allow them to be taught. At the end of the class, however, I must have covered the five primary objectives in order for the class to be successful.

You will need to start by writing a high-level outline of the course. In this case, each of the main "points" may be one of your objectives. Ask yourself, "in order to meet this objective, what do the students need to know?" Note that "know" is appropriate at the outline level, even if it is not at the objective level. Would it help your students to learn about the history of the product, or what products the competition offers, etc.? If so, put that in as part of your outline, even though it is not part of your objectives. Make sure that everything you include in your outline points to an objective. Remember this is not an agenda. You don't need to include breaks or the visit from the head of marketing. That can come later. For now, concentrate on what you believe is necessary to meet your objectives. Later, after step 5 is done, you will prioritize this even further. Once you've done this a few times, you'll be able to start this prioritization even as you make the outline.

Constructive Activities

All of the steps in the 4×8 Proficiency Design Model are important. This particular step, however, tends to be minimized in product training more than most. One reason for that may be that experts, who are the main instructors of product training, don't feel that this step is necessary for themselves, so they minimize it for others as well. To become a great facilitator, however, you must grasp the importance of creating activities that help your students construct the knowledge for themselves. Think back to when you first learned the material yourself.

The fact is that *you* don't need the exercises—because throughout your career you have completed similar exercises hundreds of times. Most of the time, you were doubtlessly unaware of your learning progress. You must force yourself to develop exercises that continue the learning, not that complete it. If you believe that you cannot be proficient in something you haven't done, you must provide opportunities, or activities, for your students to become proficient. Too often those activities are reserved for the end of the class, after the content has been delivered. However, there is an important distinction between an activity that is designed to demonstrate or assess proficiency and one that is designed to construct that proficiency.

Many experts need to change their mindset about activities. Table 9.2 shows a few of the objections I have received over doing constructive activities during product

Table 9.2 Objections for using activities.

They are time consuming	Perhaps, but they don't need to be. Create more activities that are short. An activity can be as simple as choosing the right product or solution for a very simple application—perhaps an "activity" that takes less than a minute
	Activities will take more time than merely lecturing your students. But your goal is for learning to take place, not just delivering the content
They require delivering the content first	Not always. Perhaps an assessment activity does, but not a learning activity. Many times, they can replace the delivery of the content altogether
	When content is required first, deliver the content. It is often possible, however, to create activities very early in the learning process. Doing this helps set the tone for the rest of the class
They are difficult to manage	Yes, they can be. Some will take more time than you expected and others will take less. Be sure to keep everyone involved. This can be the most difficult part and takes practice and creativity
They require too many materials	If the materials are required to learn, they are not too many. If you can reduce them in creative ways, do so, but never sacrifice learning simply out of a desire to haul fewer materials to the training event
The "good" students do all the work	Know your audience. Be very honest with your students about expectations. Make sure that all students get an opportunity to participate. If they feel that this is merely a demonstration of an ability, they may be quick to say, "I saw them do it, I get it," or "I know how to do that." Be sure to encourage all students to contribute. If that is not possible and they are working in teams, encourage the newer students to take the initiative for learning purposes
	In rare cases, your students may already be experts that can learn in the same way you do—by absorbing content. Again, the key is to know your audience

training. Whatever your personal objections may be to activity-based learning, they should be overcome by the simple fact that you are seeking to help others become proficient in your product. As a student, one must be able to learn by doing. As an instructor, you cannot know they are learning unless you see them putting that learning into action.

You should have at least one activity for each of your major objectives. How will you know that your objective was met? Don't wait for the exam at the end of the class. Instead, create activities that demonstrate that the objective is being met. If your objective was worded as an observable and measurable verb, this will be very easy. Take a look again at the objective we wrote earlier.

> "The successful student will demonstrate their understanding of network cable types by choosing the appropriate cable for a given scenario."

What activity could you create for this? One activity should be to give the student a particular scenario and have them choose the appropriate cable for that application. However, prior to that, there could be several smaller activities. For example, you could have the students sort cables by type or sort scenarios based on certain criteria that will later be used to determine cable type. There are many ways to teach that don't involve lecture. When you move from a lecture-based approach to a facilitation approach, you will rely less on a presentation tool and more and more on student activity and engagement. As your student engagement rises, so will their retention.

Once you have determined what activities you need to create learning, go back and add these activities into the outline. At this point, you can prioritize your outline, which now includes the activities. I use a simple + and – sign next to each item. If a point or topic is required to meet the objective, it gets a + sign. If not, it gets a – sign. This helps me to keep in my mind what is mandatory and what is not. The activities should almost always be marked with a + sign. If you don't, they will be the first thing to go when time gets tight, and your training will easily slide back into a lecture-based training session.

You will see that the steps that follow this one rely heavily on the importance of good activities. If the foundational philosophy and understanding that your learners will only really learn by doing is not there, the temptation will be to cut the activities at every one of the next steps. Doing so will shorten time, will allow more students, will allow greater flexibility of delivery methods ... all the signs will seem to be positive, except the one that matters the most. You will not meet your objective and real learning will not occur.

Determine Delivery Method

The sixth step is a logistical step. After you have created the appropriate objectives geared for the right audience in order to accomplish a solid business goal, and you have created activities that will help achieve those objectives, you can determine the logistics of the class. While this step is titled "Determine Delivery Method," there are actually several items that you can determine at this stage.

Delivery Method

Many times companies will want you to determine the delivery method well before you get to this point in the design process. This is when you should make that decision, however. The activities required to demonstrate that your objective was met will determine how you can deliver the training. This can, however, be influenced by other factors, which is why this step is on the same level as the activities step. Many activities can be delivered in multiple formats. Simulation and other technologies allow for significant flexibility. However, certain circumstances (as in for certification) require students to demonstrate an ability to perform an activity. A simulated activity may not be sufficient.

Duration

Now is when you can begin to make an educated guess at the time needed to deliver the training. It won't be until after the first pilot class that you can really determine how long the class will be, but, depending on how familiar you are with the topic, you may get close.

Maximum Number of Students

Another decision that can be made now is the maximum number of students that you can successfully teach at once. Like the other options, the activities will help to determine that. The number of students that can successfully complete the activities in a reasonable amount of time is likely to be the main factor in determining class size.

Other Logistics

There are other logistics that you can fill in now, such as who will qualify as an instructor, how often the class will be delivered, what the class should be called, what materials will be required and where, or what environment you can deliver the class in. When these logistics are provided at this point, instead of earlier in the process, you will be more confident that they will not change.

Level 4

Finally, you get to pack the car. As the expert, you have been waiting for the opportunity to create the content. The temptation will be to simply pull out what you have used in the past. If you do, you risk wasting the efforts in levels 1 through 3. Work on steps 7 and 8 simultaneously, regularly checking back through the other steps to make sure that they complement the rest of the design work.

Provide an Assessment to Validate the Learning

If your class is going to have an assessment of the competencies you are teaching, this is the place to begin writing those assessments. If your objectives were well written, they were both observable and measurable. You should assess based on those objectives. Ask yourself how you will know if the students have met the objectives. That will help you create better test questions.

Chapter 7 covers how to create better assessments. Here, it is sufficient to know that you can begin the process of creating the exam questions or exercises.

Create the Content

And now … drum roll … we get to content! Here is where you probably wanted to start. Instead this is where we wrap up.

Why Is Content After Assessment?

I'm often asked why content comes after, or at least contiguous to, creating the assessment. Some fear that this will lead to "teaching to the test," which is a myth in product training. Teaching to the test is only an issue when teachers or instructors are graded merely on how their students perform against a standardized exam versus on how much their students actually learn in the classroom. This should never be an issue in product training. In fact, there are at least three reasons to create the exam questions while, or even before, the content is created:

1) **It improves the exam**

   When the questions are created after the content is created, you will find that the tendency is to ask about what you delivered, not about what the students learned. Creating the exam prior to creating the content will force you to ask the question, "How will I verify that the students have learned this?" That will require developing content that gets you to the right point.

   Good exam questions determine a student's comprehension of the objective you set out to teach. If you have asked a question that adequately measures where you want your students to be, you *should* be teaching to the test!

2) **It improves the content**

   In a similar way, creating good assessment questions and projects prior to developing the content actually improves the content. Too many experts look at me with a blank stare when I ask the simple question, "How will you know they have learned it?" By creating the assessment along with the content, you will find that the answer to that question is rarely "I will tell them." Instead, you will be encouraged to create engaging ways to deliver the content. When—and only when—you know what success looks like will you be able to teach it. If you don't know how or what to measure, you haven't put enough thought into how your subject should be taught.

3) **It will happen anyway**

   Admittedly, this is a weak point, but honesty can't hurt. You were probably thinking that anyway. The truth is all teachers teach to the exam to some degree or another. If they don't, they're either a bad teacher, or they have bad exams.

In conclusion, don't worry about it. If this has been a concern of yours, work on improving your assessments, not on eliminating teaching to them. It can be very liberating, as an instructor, to know that the exam and/or practical assessment will adequately measure a student's grasp on your objectives.

Conclusion

Is what you teach important to you? If so, then you should be able to define a process that guarantees its value every time you create new training content. Following the process described above will guarantee that you get the right content to the right people and that you deliver it in an educationally sound way. Most of the steps were things you would have done anyway, though perhaps not in that order.

Whether this model completely upsets your normal content development approach or you see it as merely a common sense way to get the most out of content, try it. The learning you create will have a positive impact on your business.

Making It Practical

The 4×8 Proficiency Design Model is a simple model that helps subject matter experts create effective training. Many of the concepts apply to any good curriculum design model.

1) A clear objective uses an action verb that is both _____ and _____.

2) Knowing how activities increase proficiency, how would you respond to the objection that using activities in training is too time consuming?

3) Which of the following steps are included in the last level of the 4×8 Proficiency Design Model?
   a) Outline, constructive activities, and delivery method
   b) Objectives
   c) Assessment and content
   d) Business goal and intended audience

Before you read Chapter 10, "Pixels or Paper? How to Build the Content and Deliverables," answer these two questions.

1) What types of learning materials have you found most helpful to you as a student?

2) What is a pilot class and what are the benefits of having one?

10

Pixels or Paper?

How to Build the Content and Deliverables

When you get to the last step of the 4×8 Proficiency Design Model, you are ready to create the content for your training class. The reason for putting content at the end of your curriculum design process is to ensure that you construct it correctly. Effective training must have accurate content. If this is the first time you've used the 4×8 Proficiency Design Model, it is best to create your content from scratch. Any materials you created or used before may be useful, but starting the process from the beginning will guarantee that your new deliverables match the goals, audience, objectives, and exercises you chose for this course.

Ask the Questions Again

When I refer to building the content, I'm referring to the gathering of all the steps together into one cohesive curriculum. Start with your objectives, and then look at the outline you determined would help you meet those objectives. The outline is your guide, but perhaps not in the way you are used to. Don't just fill in each point with a new PowerPoint slide. Instead, start by asking a few questions. If you've followed the process properly, you will find that many of them have already been answered.

1) **How can a student demonstrate that they have met my learning objective?**
   This should be easy, since your objective is already a measurable and observable action verb.
2) **How can they learn by doing?**
   a) How can they practice it in the classroom or eLearning module? Hopefully some of this was already captured when you created the constructive activities in step 5. Remember step 5 is not merely about demonstrating an understanding of a topic you previously covered in a lecture. Step 5 is about actually building the learning through an activity. It is perfectly fine to have no lecture at all and teach the entire concept through the activity.

Product Training for the Technical Expert: The Art of Developing and Delivering Hands-On Learning, First Edition. Daniel W. Bixby.
© 2018 John Wiley & Sons Ltd. Published 2018 by John Wiley & Sons Ltd.
Companion website: www.wiley.com/go/Bixby

b) Related to this question, how can they demonstrate this in the classroom or eLearning module?

While learning by doing is important, it is also important to capture success. Assessing a student's ability to do certain activities may be a requirement and should have been captured in step 7.

3) **What assumptions am I making?**
This question may take a little more time to answer. We all make assumptions. Experts tend to make the wrong ones. Make sure you get some typical students to help answer this question. What do you expect them to be familiar with or already proficient in? This is a critical question to get right, since it will help fill in many of the learning gaps for your students. As you get the answers and see the gaps, fill those into your outline. Where should they be taught and how?

4) **What must I tell them?**
The very last question you should ask is, "What do I need to tell them before they begin the learning activities?" List only the information required. Get to the activities as soon as possible. Let your students learn *while* they are doing.

Asking the questions again will make sure your content is tied together with the work you've already done. Then, you can start putting the content into a tangible form that can be delivered to your students. If you have curriculum designers, technical writers, or others that specialize in this function, get their help. This book is not written to be a complete reference for instructional design, nor is it written for instructional designers. As the technical and product expert, you have a vested interest in delivering the training in the best possible format. You want people to remember what you taught them well after the event. Asking the questions again can help.

Create a Student Guide

After you have asked the questions again, you can start creating a student guide.

Wait … before the PowerPoint?
Yes, before the PowerPoint.

If your class is more than a couple of hours long, you need to give something to your students to take with them—something that was designed specifically for that, not just a printout of your slides.

The ability to quickly and easily show content on a screen to all of your students at once is a wonderful technological marvel and an amazing teaching tool. But it is just a tool, and you shouldn't use a hammer to tighten a screw.

Create the student guide first, because you are more interested in what your students learn than what you teach. Handing them a simple printout of your own notes signals that your primary interest is in delivering the content. Providing them a learning guide designed to help them learn in ways other than your lecture tells them that your primary concern is that they learn.

A learning guide is part tourist guide and part scrapbook. It is a map with highlights and extra information of what you will be seeing and a place to hold reflections and pictures of what you visited. It is both a syllabus and a workbook.

I will use the terms student guide and workbook interchangeably. Here are some key elements that are important to include in a student guide:

1) **Whitespace**
   Space for writing is number one for a reason. If students are going to take notes in your workbook, it must be created for that. In addition, whitespace will make the workbook look neat and professional and give credibility to your class.

2) **The objectives**
   Always state the class objectives at the beginning of the student guide. Instead of a bullet point or numbers, I like to use a little box next to them. At the end of every class I go back to the objectives and ask the students to check the box if they felt that objective was met. This gives them a reference six months from now when they need that reminder.

3) **The outline**
   This may seem obvious, but always include the outline of the subject matter being taught, even if the format is different. You don't have to have the numbers and letters, but you do need to have the headers.

4) **Questions or exercises**
   The student workbook is another way to engage your students. Feel free to insert questions, even some that you may not ask in class. The point is to augment their knowledge, not to examine them or create busy work for them.

5) **Extra insight**
   You can never teach everything. Use the student guide to encourage learning beyond the classroom. Adding extra information to your student guide increases the value of your class. Highlight it with an indicator so students know it is information that will not be covered in the objectives.

6) **Keys and place indicators**
   It is a good practice to include visual indicators of some sort for the student. Only include a key if the visuals aren't obvious. You aren't writing a technical manual, but a glorified note-taking workbook. Use page or slide indicators to help students keep up, since you aren't giving them the actual slides.

7) **Graphics**
   The graphics you use should match what you use on your slides. Make sure you have the appropriate permissions to use them. You don't need to use all of them, of course. One of the benefits of a student guide is that it frees the instructor up to use even more graphics in their presentation.

8) **Copyright information and indemnification**
   Lastly, always include the copyright information in the student guide. Make sure you don't put anything in the document that you don't have permission to use. Include a notice that the document is intended for use by students only. You may want to include a statement of indemnification, depending on what you are training on. For these items, I would seek help from your technical writing department and legal office.

Create Your Visual Aids

Now you can make those PowerPoint slides. At least, that is what most people think of when they think of facilitation visual aids.

Creating Presentation Slides

A quick disclaimer. This is not a book about creating good presentations. There are many great resources for that and I encourage you to check them out.

Here are a few tips that will help to facilitate effective learning.

Use the Software Correctly

Although there are other sources of presentation software, such as Keynote and Prezi, I will refer to PowerPoint, since it is the most widely used tool. If you use something different, replace PowerPoint with whatever tool you use. They are all merely tools to help communicate and create engagement. When the tool you are using starts to decrease engagement, it is being overused.

Someone has said that PowerPoint is the worst thing that ever happened to training. I understand what they mean. You cannot, however, blame the tool for poor training. Someone very well could have said that about the chalkboard at one time. Use the right tool for the right reasons and in the right way.

1) **PowerPoint is not a book**
   Don't write it like one. You don't need full paragraphs of text, or even full sentences. Keep your text size large enough to be read by everyone and short enough to be read at a glance. If you need longer explanations, quotations, formulas, or charts, put them into the student guide or print them as a handout.
2) **Use less text, more graphics**
   Most technical experts put way too much text on the screen. There are several reasons why they do this, but the most obvious is convenience. They feel they can create one document that will serve as a presentation tool and technical manual. It doesn't work well for either.

   Too often, instructors who watch videos of themselves are surprised to see how much they face the screen while teaching. Many times, that is because they've written such a long sentence on the screen, they have to look at it to read it. Help yourself out by putting less on the screen. This is one case where less truly is more.
3) **Use presenter notes**
   If you really need to put more text in the slide, put it in the notes.
4) **Use presenter mode, or something similar**
   Using this mode doesn't show just a duplicate of your screen. It allows you to see the next slide and your speaker notes as well.
5) **Learn the shortcuts**
   All software programs have shortcuts. Learn how to use them to blank the screen ("b" while in presenter mode in PowerPoint), white out the screen ("w" while in presenter mode in PowerPoint), and so on.

Don't Rely on a Presentation

If you rely too heavily on your slides, losing them due to technical reasons can ruin your training. That is another benefit of a student guide. No slides? No problem. Here are some ways to keep a technology glitch from ruining your training class.

1) **Save a copy to a portable drive**
   If your computer chooses the worst possible time to update or crash, you can borrow another. Also, many newer projectors have a USB input that allows one to connect a flash drive and present directly from the projector.

2) **Always have a backup plan**
   You never know when the bulb will go out, or your computer will decide to quit. Have a plan B. Generally, your student guide or textbook is your backup plan. However, you may have equipment that you need to gather around, pictures you may need to distribute, or other ways to teach without a projector and screen.

Don't Let the Presentation Tie You Down

In a similar way to the above statement, instructors often rely too much on the order of slides and forget that they are teaching individuals with individual needs.

1) **Use hidden slides to increase flexibility**
   One nice feature in PowerPoint and Keynote is the ability to hide slides. Your presentation will skip hidden slides unless you click on them directly. You can put in some coffee break slides, but keep them hidden. When you see that the class needs a short break, click on the slide and will look like it was planned all along. This keeps the student that yawned from getting embarrassed!

2) **Use a remote presentation tool**
   A remote presentation tool frees you up to concentrate on your students. Learn how to use it well. It does more than just advance the slides. Become familiar with the other functions as well.

   A good example is the "mute screen" button (most remote presenters have one), which blanks out the slide. You'll find that whenever the slide disappears, students' eyes immediately go from the screen to you, the instructor. I use this frequently. One way for you to use this is when you have asked a question and want to make sure you have the students' attention. Blank the screen. When a student asks a question, you can also blank the screen out of respect for them. It will pull the attention away from the screen and toward the student asking the question, signaling that you take questions seriously and want the entire class to learn from them.

   In order to learn how to use your tool, you will need to practice using it. Students should hardly know it is in your hand. Modern remotes don't require that you point it at the screen or computer, and giving it a good shake when you change the slide doesn't help, either! You may chuckle, but I see that all the time.

Know Your Material

1) **Know what is coming next**
   The more you practice, the more comfortable you will be. You don't have to know every word, but you do need to know the flow of your material.

   You don't need to use line-by-line animation. Often, the text is in the student guide anyway, so you are not surprising them in any way. If the order is not important, you don't want to be forced into a particular order by having the text appear point by point.

2) **Don't make excuses**

One of the most uncomfortable feelings as a student is to watch an instructor struggle. Most of the time, however, students wouldn't have known the instructor was struggling if he or she hadn't informed them that they were. The fear of a few seconds of silence has forced many new instructors to make excuses they didn't need to make.

Never say "I've never seen this before" or "these aren't my slides." If you must, explain that to the students one time, at the beginning, but not when you don't understand the slide. Instead, take a few seconds to absorb the slide yourself. Those few seconds may feel like a long time to you, but they will not to your students. Most of the time, you'll be able to collect your thoughts and continue, without making an excuse.

Creating Handouts

I like handouts. Handouts can be anything from a chart or graph, an exercise, or anything that you don't want to include in the student guide that students get at the beginning of class. Here are some reasons to use handouts and things to consider when you do.

Use a handout…

1) If you have a worksheet or other material you want to collect.
2) When you don't want them looking at it prior to covering the material.

Students will look over the student guide right away. Usually, that is a good thing. When they know what is coming, they won't waste your time asking about it, and adults learn better when they are prepared in advance, even if that just means seeing it in the outline.

There are times, however, when you don't want to divulge certain things before it is time. Use a handout instead.
3) When the detail is too long or large to fit in your workbook.

I don't like to use multiple paragraphs, even in my workbook. If one requires that, I create it separately. Formatting and continuity is important to me.
4) When it is from a third-party source.

Get permission to copy it, but keep it separate from your learning guide.
5) When it is part of an exercise.
6) When it is supplemental material.

Statement of Indemnification

It is a good idea on any training material to include a statement of indemnification. Seek the advice of your legal team. Let them write a statement, or give them one to modify. The statement is not a foolproof guarantee against lawsuits, but it can help to protect you if someone takes your training and then has a failure that leads to any sort of loss. All of your documentation should include a statement similar to the one in Figure 10.1.

Figure 10.1 Example indemnification statement.

> **[Example only — seek legal advice]**
>
> This document is for training purposes only.
>
> While every effort to maintain relevant, current and accurate information has been made by the training staff, all technical and other information should be drawn from the applicable technical manual or revision release notes. By using this document you are accepting these terms and agreeing to indemnify all [company name] personnel from all liability for any loss, damage, claim, or demands therefore, on account of any equipment or system failure, whether caused by the information written, stated, implied or otherwise.

Create an Instructor's Guide

This step may be optional, depending on the type of training you are giving. If more than one instructor will be teaching the class, an instructor's guide is advisable. You may also be able to use the speaker notes in your presentation software.

If you do choose to create an instructor's guide, here are some things to consider:

1) **Keep it as similar to the student guide as possible**
   The key here is to be able to tell exactly where the students should be. It gets frustrating to students when you say, "I'm on page 13, what page are you on?"
2) **Don't give verbatim statements**
   Too many instructor guides tell the instructor exactly what to say. Give ideas, but not exact wording. Giving too much implies that the instructor can just read their guide as they "teach" it for the first time. Not much will actually be taught.
3) **As much as possible, make the instructor's guide a live document**
   Every time you teach your class you should make it better than the last one. A good instructor's guide provides a place to keep those notes and ideas, both for you and for others that may teach the class. It should constantly be added to and revised, even if the curriculum itself is not.

An instructor's guide does not make a good instructor. Having an instructor's guide is optional, but having a pilot class is not.

Running a Pilot Class

There are two types of pilot classes. In one class, you are piloting the curriculum, and in another you are piloting the instructor. Occasionally, both will occur at the same time. Hopefully, that doesn't happen often.

When an Instructor Teaches This Class for the First Time

Everyone has to start somewhere. Don't be afraid to state that, unless there is a good reason not to. Never teach the class alone the first time. If the class is long and has multiple modules, the best scenario is to co-teach a few times before going solo. When you do, always seek feedback so that you can feel more comfortable the next time.

Teaching a pilot class does not mean that this is the first time you have ever taught. However, you should teach a pilot class whenever it is the first time to teach *a new* class or curriculum.

When This Class Is Being Taught for the First Time

Another type of pilot class is when the curriculum is being taught for the first time. In this case, what is being tested is whether the course does what it says it does and how long it will really take to do that. This is a chance to try out new exercises and delivery methods as well.

Regardless of whether the instructor is teaching the class for the first time, or the curriculum is being taught by anyone for the first time, there are a few things they should consider when running a pilot, or beta, class.

Handpick the Audience

Never open up a class to the general public without first running a pilot with an audience willing to provide feedback and insight. Be honest with everyone that this is a trial run and solicit their help to improve future classes. A pilot class should be comprised of the following:

1) **A notetaker**
   Someone should be designated to take detailed notes about how things go. Everything from timing issues to questions to equipment issues should be noted. Both positive and negative aspects of the class should be noted. Don't leave this up to the instructor. It is difficult to teach and take these types of notes at the same time.
2) **Some from the target audience**
   If possible, get some "friendlies" from the target audience. This will help to make sure you are offering them what they need. Be very clear to them that the class is a pilot class and that you expect their feedback. Give them an incentive to help you make the class better, or they may feel like they got the raw end of the deal.
3) At least one expert other than the instructor.
4) At least one stakeholder.

Plan on Extra Time

If you think the class will take 2 days, schedule two and a half. This will allow for debriefing if there is extra time. Remember part of the reason for running a pilot class is to determine how much time it takes to teach your curriculum. Up until now, everything has been a guess. Give yourself more time than you think. Trust me, you'll be glad you did.

Be Aware of Too Many Auditors

I have run pilot classes in the past that had high visibility. I started getting multiple requests for visitors and auditors. They all promised, of course, to sit quietly in the back and be virtually unseen. Be careful about allowing too much of that. Too many opinions will change the class, no matter how "unseen" they think they are.

Debrief with Everyone
If students were willing to come to your pilot class, the least you can do is give them the courtesy of getting their feedback. This should include more than a standard survey. Find out what they liked and would keep and what they think could have been better. You don't have to use all of their ideas, but you do need to listen to them. Some of the better ideas I have had for training classes have come during these sessions. Your note-taker should take notes during this as well to prepare for the real debriefing, which comes next.

Debrief with Your Core Team
A second level of debriefing is to go over the notes taken and make the necessary changes. You must put egos aside for this meeting. If the participants can't be honest, the class won't be its best. This can take some time. You may want to schedule a separate meeting for each objective just to make sure you are thorough. Spending time here will save time down the road. It also generates support, if the right people are involved.

Conclusion

Can you identify what content should be in a student guide and what should be shown on the screen? Can you use your visual aids effectively?

When you provide more than one way for your students to absorb your material, you are increasing the chances that they will retain it and use it after class. This doesn't mean that student guides must be printed on paper, in spite of the catchy chapter title. Student guides can be created for digital tablets as well. The main thing is to make sure they are being used for their intended purpose. It is better to use the student guide as your teaching reference than to use slides as a student guide.

When you put some thought into the structure of your class and when you have run a pilot to make sure students can improve their proficiency in your class, you will be ready to deliver a professional training class.

Making It Practical

Creating both learning materials and presentation materials takes time and must be valuable. Take some time to reflect on what you can do differently in your training classes.

1) What values, if any, do you see in providing students with a student guide?

2) How might your presentation look differently if you change it from being a student guide to a visual aid?

3) How will you teach your next class if there is no electrical power?

Review of Part Three: The Structure of Hands-On Learning

1) Consider the design of a product training course, including the creation of visual aids, student guides, and handouts, as laid out in Chapters 8–10. Is there anything about putting content last that makes designing the training easier? Is there anything that makes it harder for you?

2) What is the main thing about the structure of your product training that you want to do differently as a result of reading Chapters 8–10?

Part IV

The Facilitation of Hands-On Learning

11

Speak Up

Effective Verbal Engagement

Your presence in front of your students is a key part of your teaching success. The skills you need to acquire will take practice and patience. There is no magic formula that will turn you into a great instructor overnight. The important thing is to improve in at least one area every time you have the opportunity to teach. When you do, you will learn—by doing. After all, facilitating technical training is a skill that requires proficiency. You will learn just like your students, except that they are learning about your product solutions, and you are getting better at teaching them.

In Chapter 7 I differentiated between a presentation and a training class. In Chapter 8 I encouraged you to focus on becoming a facilitator, instead of an expert. There are many other comparisons I could make here that would help to differentiate between an expert's natural tendency when in front of subject matter beginners and what they need to be. If your LinkedIn page is anything like mine, you have seen many different memes about the difference between a manager and a leader. While a few wording changes may be required, the spirit of those memes resonates in a learning environment as well. Those who prefer to "manage" will tend to be natural lecturers and presenters, while those who prefer to "lead" will more naturally gravitate toward facilitating.

In the next three chapters, I will pull some of those contrasts together in a practical discussion about how to do more facilitating and less presenting. Good facilitators engage their students, inspiring in them a desire to internalize what they are learning. It starts by being a good communicator in the classroom. The most obvious way you communicate to your students is by speaking to them. The verbal skills you must acquire are numerous and may seem daunting. Good news—most are just common sense. The difficulty comes in applying common sense when the job you are doing is an uncommon job.

I've broken the sub-skills required to be a good verbal communicator into ten areas; nine of which I will cover in this chapter. The fear of silence is so important that I will cover that separately in the next chapter. The skills are divided into two main groups. The first group of skills have nothing to do with what is actually stated, but everything to do with how it is received by the listener. These are skills not restricted to speakers. They apply to musicians as well, since they are the way we color, or decorate, our words to give them meaning. I refer to them as *decorative speaking*. The second set applies to what is actually stated and is most challenging for

Product Training for the Technical Expert: The Art of Developing and Delivering Hands-On Learning, First Edition. Daniel W. Bixby.
© 2018 John Wiley & Sons Ltd. Published 2018 by John Wiley & Sons Ltd.
Companion website: www.wiley.com/go/Bixby

those who must improvise or teach in a setting where you do not have the opportunity or desire to rehearse every word or memorize your speech—this skillset I refer to as *declarative speaking.*

Decorative Speaking

What makes music interesting and enjoyable is the variety of ways the sounds are decorated. Some of you would prefer to listen to hard rock, while others would love a classical concerto. We would all agree, however, that if musicians don't decorate the music—if it is monotone, or consisting of one tone, with no variety—the music is not enjoyable to listen to.

The same is true when you are speaking. In many ways, speaking differs very little from singing. I am a choir director. Often I find myself teaching my choir how to "speak" certain words or sounds. Here, I'm doing just the opposite. I'm teaching speakers how to "sing." It is no secret that taking voice lessons is helpful to improve your public speaking ability and that many singers are also good public speakers.

What makes singers better speakers is less about their musical abilities and more that they have learned to control what sounds they want the audience to hear. Even if you have no musical ability at all, you can learn to control six things that good singers are in control of at all times. When you do, you will find it will dramatically improve your speaking ability. Those six things are energy (or emotion), breathing, pitch, tempo, volume, and articulation.

Controlled Energy

I'll start with the one that is the hardest for me. It is not unusual for me to go to the restroom prior to starting a training class simply to pump my fists, slap my cheeks, and tell myself to get energetic, almost like a boxer dancing around before a big fight! As the instructor, I am well aware that the beginning of a class is critical in setting the tone for the rest of the training.

Musicians are often more energized by the music than they are the crowds. You might be surprised to know how many of them are actually introverts. I happen to be an introvert as well. When I have the opportunity to teach about adult learning, I get energized by the topic. Often, when someone thanks me for my obvious passion and enthusiasm for teaching adults, I smile inwardly, knowing that I will crash, exhausted when I get home.

You may be different. You may need to calm yourself and tone down your energy before starting the class. The key is control and knowing what is best for you.

Too much energy can lead to speaking too fast, too loudly, or not listening well, all things I'll cover later in this chapter. Too little energy will rub off quickly on your students. They will never have more energy, or be more eager to learn, than you are to teach them.

Controlled Breathing

Many speakers never think about controlling how or when they breathe. This may sound overly simplistic, but air is important. You will not be able to control your pitch and volume without first controlling your breathing. I have helped many people overcome their nerves simply by taking the focus off of nerves and focusing on breath control. I don't even tell them it is to help their nerves. In a way, that is merely a secondary benefit, though it is often the more appreciated one!

The way to start controlling your breathing is to make sure you are breathing properly. Many people breathe differently when they are nervous than they do when they are at rest. Try this simple exercise.

Normal Breathing

Lie down flat on the floor, on your back. Place this book on your stomach. Breathe normally. Notice that when you inhale, the book rises and when you exhale, the book lowers. Now stand up. Place one hand on your chest and the other hand over your stomach. Breathe deeply from your lower lungs. Imagine you are breathing out your toes, if you must. Make sure your lower hand mimics what the book did when you were lying down. That is how you want to breathe most of the time when you are teaching.

What you are doing is breathing from your diaphragm, a muscle under your lungs that pulls the air down. You are filling your lungs to their total capacity. If you are nervous, however, you will likely do just the opposite—or breathe from your chest. That is because your chest is closer to your mouth and you can get the air there more quickly. Your brain knows you need air and takes over unless you deliberately tell your brain that all is okay, you are still in charge. Breathing quickly from the chest is what is often referred to as hyperventilating, something that won't happen if you are breathing from your diaphragm.

Controlled Pitch

You may wonder why pitch is important in speaking. You are correct to assume it is not equal to the importance of pitch in music, where precision and consistency are paramount. Speaking pitch is much less precise, but only slightly less important.

The goal here is not to alter your natural speaking range significantly. It may, however, be necessary to stretch it in one or both directions. You may not realize the range of pitch you already have. Try this fun experiment, one I borrow from the game, Moods, by Hasbro.

Exercise 11.1 The influence of pitch on our words

Pretend you are a boss, speaking to one of your employees. Choose one of the following "moods" and then read the phrase below.

1) Arrogant
2) Happy
3) Angry
4) Serious
5) Indifferent
6) Sneaky

"Come see me in my office."

Notice that your pitch changes with the different moods. As you get more excited, your pitch tends to rise. "Indifferent" was likely toward the lower end of your normal range. And why isn't there a "normal" mood? Because you are always communicating something. Either you are excited about what you are teaching them, or you are indifferent, or some range of other emotion, but you are not communicating "normal."

But why does this matter to trainers? Because often times, even with technical training, what you say is less important than how you say it. Communicating an excitement and confidence is crucial, both for the success of your product and for your success as an instructor.

Another thing to remember about pitch is to maintain the pitch energy through your entire thought, sentence, or phrase. Many new speakers start a sentence with energy and end it in a fizzle. The end of your sentence is just as important as the beginning. Make sure it sounds that way!

Controlled Tempo

Another crucial area of control is tempo. Unlike musicians, who have rhythms and accompanying instruments to keep their tempos in check, speakers must control their own tempo. Most trainers speak too fast. David Lewis and G. Riley Mills use techniques used by professional performers to help business people (and trainers) learn to communicate better. In their seminars and in their book, *The Pin Drop Principle*, they suggest that most of the well-known speeches average around 125–150 words per minute. They suggest an exercise using the first 125 words of Lincoln's Gettysburg Address to find out what your average tempo is. I have used that exercise a number of times and have found that most students complete the first 125 words well before the 1-minute mark, usually around 45–50 seconds.

The key is to control your tempo. Earlier, I suggested that you need a lot of energy. Often, however, we tend to speed up when we expend more energy, so it takes more control to keep the tempo down. If you are nervous, you will also tend to speed up, almost as if your brain wants to be done more quickly. It is amazing how slowing down a little can give the students the feeling that you are in control. You may still feel nervous, but you won't sound nervous, simply because you've reduced your tempo.

If you are one of the very few who speaks too slowly, work on controlling a slightly faster tempo. You still need to be you. Don't lose your authenticity while employing any of these techniques. Also, remember that this is an average tempo across a longer period of time than just 1 minute and it includes pauses between thoughts. In Martin Luther King Jr.'s famous "I Have a Dream" speech, he has portions that were delivered at a much slower rate than 125 words per minute and portions toward the end that were much faster. The average, however, came out at right around 119 words per minute.[1]

Controlled Volume

If speakers tend to speak too fast, they also tend to speak too quietly. Most trainers are not a good judge of what their volume should be. It should not be difficult, however, once you shed the idea of merely giving a lecture and begin to facilitate the learning as a two-way conversation. It will be much more obvious if your students are not hearing you well when you are asking questions and inviting feedback on a regular basis.

As with the other elements, the key here is both control and variety. The issue is not a matter of *sometimes* speaking too quietly, or *sometimes* raising your volume. Problems arise when students must consistently strain to hear you or feel as if you are constantly shouting at them. Variety is good. Consistent extremes are not.

If you are teaching in a classroom arrangement with multiple rows of seating, verify that from the back row they can hear you well. One of the main reasons I like the U-shaped classroom is because I can get equally close to everyone.

For a number of years, I worked for a professional audio company and had the opportunity to work alongside some of the best audio engineers in the industry. One thing I learned was the direct correlation between hearing and fatigue. When a student has to strain to hear, their brain must concentrate on hearing instead of understanding and it gets tired faster.

That is more than just an interesting bit of information. According to one US government source, about 13% of adults have hearing loss in both ears.[2] That means that if you have even a small class, you are likely to have at least one student with hearing loss. It is not inappropriate to speak just a little louder when speaking to a group of people. Combining a slightly louder speaking volume with regular eye contact will increase retention and reduce fatigue and disinterest in your students, and students with hearing loss will be grateful.

Controlled Articulation

Articulating your words well will also help your students' brains to focus on learning the concepts you are teaching instead of trying to decipher the words themselves. This is especially important when teaching students in a multicultural setting or those for whom English is not their first language.

The key to all of these musical elements is control and variety. Too much volume all the time can be a drag. I like to deliberately get quiet when I want to get everyone's attention to say something very important. The key word there is "deliberately." Even over-articulating can make you sound fake as if you're seeking to sound superior to your students. Make sure you get feedback from a trusted source or record yourself so you can know where you need to improve.

Declarative Speaking

Impromptu speakers have some things to perfect that vocal musicians do not. Because their music is prewritten, they don't have to worry about things like jargon, verbal crutches, or poor grammar. Okay, maybe their songs use poor grammar, but that is the writer's fault, not their own. These few items refer to what is actually stated and why it is important to get it right. This requires controlling jargon, not using unnecessary words, and not allowing poor grammar to be a distraction.

Controlled Jargon

Every industry has its own set of jargon. It is impossible to get rid of it completely. Nor would you want to. If someone is seeking to become an expert in your industry, the language of the industry must come with it. But it must be controlled. Use it, but use it deliberately. Your audience may include both veterans of your industry and novices who have not yet learned the jargon. You should be especially aware of acronyms and terms you use specifically in your company.

There are two important things to avoid when tempted to use jargon. First, avoid assuming your students know it. Always explain the term. Don't ask them if they know it, just do it. Second, don't use jargon to impress your students. You don't need to impress them. You're the instructor; they're already impressed.

Verbal Crutches

Singers are lucky. They don't need verbal crutches. Verbal crutches are extra words speakers insert into their sentences that do not provide any value to the sentence. "Umm" and "ah" gets all the bad publicity, but umm is not alone.

The main reason we throw in extra words like umm and ah are because our mind and our mouths are at two different points in the conversation. You rarely think about the actual word you are speaking right now. Instead, your mind is somewhere ahead, trying to navigate the dictionary for words further along in the sentence. When the words you are speaking start to catch up and get uncomfortably close to where your mind is fumbling around for the right word, you throw in an extra word or catchy phrase.

Occasionally, a crutch can start out of good intentions and then turn into a bad habit. For example, I struggled with ending many of my thoughts with the same phrase/question, "Do you understand what I mean?" In the beginning my intentions were good. I was trying to include the students and verify their understanding. But there are better ways to engage them, rather than a habitual question with an expected "yes" for an answer.

So what is the answer to getting rid of your verbal crutches? First, admit that you have them. Almost all of us do. Those that don't have likely worked hard to rid themselves of them and probably continue to do so on a regular basis. Second, understand why you use them. While most verbal crutches start as an attempt to buy time for your brain, they can turn into a habit. You might try the following nonscientific exercise. It may help to determine how to address the problem.

Exercise 11.2   Counting your verbal crutches

Record yourself speaking about a topic of your choice for 3–5 minutes. Replay the recording and count the word crutches. Now answer the following questions:

1) How many verbal crutches did you count? Give one for each verbal crutch up to 10.
2) How well do you know the topic? On a scale of 1–10, 1 being completely new to it and 10 being an expert.
3) How many times have you given this lecture/speech/training? Give 1 for each time you've delivered it up to 10.

If you scored 3–22, the verbal crutches are likely a communication issue. If you scored 23–30, it is likely a bad habit.

If Your Use of Verbal Crutches Is a Communication Issue

Knowing that the use of verbal crutches stems from a communication gap between two simultaneous streams of information—one between your brain and your mouth and one between your mouth and your listeners—will help you to develop a solution. The most common and successful solution is to simply replace the verbal crutch with silence. Small amounts of silence (try timing how long it takes you to say "umm!") are actually appealing to the human ear, and your students are not likely to detect the silence, much less be annoyed. All of them would prefer silence to a repetitious "umm," or any other word.

The second way is to become extremely familiar with your topic—and not just the topic, but the structure of the training you are delivering. Take advantage of natural breaks to look ahead and be prepared for what is coming. The first few times you teach a class you may only be able to look at the next topic or agenda item, but if you deliberately stretch yourself, you will find that you can look further and further ahead. If you combine that with a deliberate practice of pausing (even for a split second) when you want to say a crutch word, you will find that you can reduce the crutch words significantly in a short time.

If Your Use of Verbal Crutches Is a Habit

If a verbal crutch has become a habit, or just part of your regular speech, the most important thing is awareness. Most people are too polite to point out our verbal faults, so find a friend or confidant you can trust to help with this. Create practice sessions where you talk about a particular topic for 3–5 minutes and allow them to record you or count your crutch words. The approach should be similar to physical exercise. Look for small gains over a period of time; don't try to conquer verbal crutches all at once. Beware of simply replacing the word or phrase with another one. The only thing you want to replace unnecessary words with is silence. If your city has an organization like Toastmasters, I encourage participation. It will help cure this habit.

Poor Grammar

The main goal in eliminating jargon and verbal crutches is the same one underlying the goal of using correct grammar—avoiding any distractions to the learning process. No one expects you to be a grammar instructor (unless you are!), and we all make mistakes. I will not list here all the mistakes that could be made—this is not a grammar book. What I do want to emphasize is the importance of eliminating egregious errors that will interfere with the message you are trying to deliver. If you struggle with certain grammar issues, don't try to correct all of them at once. Pick one or two and work to correct them.

Conclusion

You can quickly learn to control these nine (and soon to be ten) items. The key is to work on one of them at a time. The approach I mentioned for verbal crutches works for each of these areas. Find someone who you trust, tell them which area you are working on, and ask them to help you. Being aware of what your strengths and weaknesses are is important. A great way to see that for yourself is to record yourself teaching a class. Even just 10–15 minutes is probably enough to detect which things you need to work on. Use that video to fill out the self-evaluation sheet on the next page. Even if you use a live class to record yourself for future learning, don't approach the class as a practice session. Practice should happen outside of the performance hall. And don't forget that practice doesn't make perfect; it makes permanent. Be careful to practice the right things and get regular feedback. Only then can your practice pay off in a perfect performance.

Making It Practical

What you tell your students is important. The way you deliver the message is also important. Even experienced public speakers need to make sure they stay in control of both their declarative and decorative speech.

1) What do you think is the difference between decorative speaking and declarative speaking?

2) Complete the worksheet shown in Table 11.1.

Table 11.1 Verbal communication chart.

		Verbal communication	An area of strength	An area to improve
		Without sharing information, mark three areas you believe are a strength of yours and three that you feel you need to improve most in. Ask a friend or coworker to do the same. Compare your results.		
Decorative	Energy	**Your students will not be more excited to learn than you are to teach.** Demonstrate an enthusiasm and passion that inspires your students to take the learning beyond the classroom.		
	Breathing	**The best way to control nerves is to breathe properly.** Demonstrate control of your nerves and related influences, like tempo and word crutches as a result of controlling your breathing.		
	Pitch	**What you say is often less important than how you say it.** Demonstrate a variety of tone and interesting speech that is colorful and pleasant to listen to and demonstrates communication through more than just the words that are spoken.		
	Tempo	**Learning happens at the rate of absorption.** Demonstrate an ability to control your tempo and to use pauses, questions, and other absorption techniques in a facilitation environment.		
	Volume	**Students cannot focus on learning when they are struggling to hear.** Demonstrate an ability to speak at a volume that makes learning easy and to use a variety of volumes to generate interest and control in the classroom.		
	Articulation	**Students cannot learn if they cannot understand.** Demonstrate an ability to enunciate words so that students understand clearly and accurately the first time.		

Table 11.1 (Continued)

			An area of strength	An area to improve
Declarative	Jargon	**Jargon adds unclear language that disrupts learning.** Demonstrate the courtesy of defining acronyms and other industry jargon the first time you use it. Don't assume that your students already know it.		
	Grammar	**If it is worth saying, it is worth saying correctly.** Demonstrate a sensitivity toward using correct even if imperfect grammar. The goal is to be understood without being a distraction.		
	Word crutches	**Unnecessary words are unnecessary.** Demonstrate an ability to remove clutter so students absorb information quickly.		
	Silence	**Perhaps the most unused communication tool is silence.** Demonstrate a willingness to embrace silence as a teaching tool. If it is worth teaching, it is worth giving them the time to absorb it.		

Before you read Chapter 12, "Shut Up: Effective Listening and Engagement," answer these two questions.

1) Explain why you think instructors need to be better listeners.

2) Describe your opinion about the following statement: "You can't engage your students if you aren't listening to them."

Notes

1 Lewis, David and Mills, G. Riley. The Pin Drop Principle. San Francisco, CA: Jossey-Bass, 2012.
2 National Institute on Deafness and Other Communication Disorders. https://www.nidcd.nih.gov/health/statistics/quick-statistics-hearing. Quoted by my friend and advocate, Gabriella Broady, here: https://www.linkedin.com/pulse/say-what-gabriella-broady-m-ed-?trk=mp-reader-card (accessed August 8, 2017).

12

Shut Up

Effective Listening and Engagement

You can be a great instructor. You don't have to be an accomplished public speaker to be an excellent facilitator. In fact, of the many challenges instructors face when speaking, one of the hardest things for new facilitators to do is to *stop* talking. However, because proficiency training is about developing skills and because those skills are developed by individuals, lecturing should be minimized, and listening (by the instructor) must be maximized. Listening, in a facilitation environment, involves more than just ears; it also involves eyes. Listening to and observing your students is a critical part of helping them become proficient with your product. But to do that, sometimes you must shut up. You must let them talk and let them perform and you must become an auditor of their learning progress.

Obviously, you must talk to your students. But your goal is not to become a great speaker. Even though practicing verbal skills is important, your ultimate goal is to get out of the way. You need to perfect speaking, not so that you can have a great delivery, but so that you can avoid being a distraction from the learning. Like the sound technician that does the job so flawlessly that not one microphone squeaked, not one level setting was too high or too low, and most concert-goers were unaware of his presence— it is going to take work. You don't need to impress your students with your great speaking skills, but you can be successful if your students leave your class forgetting that you spoke at all. It is going to require listening and knowing what to listen for.

What You Are Listening for

There are many things you can find out if you make listening a deliberate part of your training plan. Engaging is a form of listening, and engaging your students is important— not only because the continual drone of one voice may put them to sleep but also because it actually increases learning. There are three main reasons why it is essential to frequently shut up and listen to our students.

First, since proficiency is built on experience, knowing the experiences that students bring to the classroom can be very helpful. Second, you can readily discern if they are ready to learn and apply what you are teaching them. Finally, listening is necessary in order to find out what your students are learning in your class.

Product Training for the Technical Expert: The Art of Developing and Delivering Hands-On Learning, First Edition. Daniel W. Bixby.
© 2018 John Wiley & Sons Ltd. Published 2018 by John Wiley & Sons Ltd.
Companion website: www.wiley.com/go/Bixby

What They Already Know (or Think They Know)

The *science* of effective facilitation is that learning new knowledge requires past knowledge, and acquiring a new skill requires past skill. The *art* of facilitation is taking the variety of past experiences and turning them into a cohesive learning experience for everyone. To do that, you must know what experiences and knowledge your students are bringing with them to the classroom.

The principle of listening to what your students already know applies to a short workshop as well as much longer classes. In a very short seminar or workshop, you may choose to ask a simple question or two, perhaps by a show of hands, while in a longer class you need to address each objective separately and in multiple ways throughout the class. Either way, listening is vital if you are going to be effective. Listen to their questions, not just to provide an answer, but to determine the experience behind the question being asked. If possible, provide exercises that not only demonstrate learning of the lesson you are currently teaching them but allow for some demonstration of past learning as well.

Unlike a presentation or speech to hundreds of people, a product solution training class is trying to help each individual improve their proficiency in some aspect of your product. A presentation may be considered successful if only a small percentage of the listeners are persuaded or inspired by the performance. It is possible, however, for a training class to be considered unsuccessful even if only one student fails to improve their proficiency. To ensure that you obtain the desired result, you must know what experiences all of your students are building on. To do that, you must find ways to listen to them.

What They Want to Learn

Another reason to listen to your students is to find out what they want to learn. As I covered in Chapter 4, adult students must take ownership of their learning. If they want to learn, they will learn. Sometimes, however, they want to learn something different than what you intend to teach them. This usually happens when courses are designed as a "one-size-fits-all" course—an approach all too common in expert-driven training.

The best learning will happen when the learner's desire to learn is aligned with your objectives. If it isn't, you need to do what you can to make that happen. If you don't listen to them, you take the risk of teaching students who aren't interested in what you are teaching them, or—and this could be worse—wasting time trying to align objectives when they don't need to be.

Your students want you to listen to them. They want to know if you are going to teach them what they want to learn. If you simply ask them why they are at the class, you will get a variety of answers. Sometimes their objective will be very broad—they want to "learn more" about your product. The adage that students "don't know what they don't know" applies here. Other times, they may be testing your knowledge—they want to know something very specific that may go beyond your intended objectives. Either way, spending a little bit of time listening to their objectives for being in the class saves you from wasting time later.

When you listen to your students, you gain credibility as a true subject matter expert. Only someone comfortable with the topic they are teaching will be brave enough to make themselves vulnerable to their students and honest about their limitations. Your students don't care about how much you know; they care about how much you can help them learn. Letting your students tell you what they want you to teach them creates a communication link that will immediately improve your effectiveness as an instructor.

What They Have Learned

A third reason for you to shut up and listen to your students as you train them is that you can more readily determine what they have learned in the class. As a proficiency instructor, your job is to communicate, getting regular feedback to make sure your students are connecting with the information you are delivering and that they are able to internalize it for themselves. Without the listening aspect, you will not know if they are ready for more information or if you need to find other ways to solidify what you are teaching before you move on. Otherwise, your training becomes purely a lecture without any verified results.

While giving an exam or quiz at the end of the class is certainly a form of listening to what your students have learned, exams do not always provide the opportunity to address any learning gaps that may exist. I suggest asking questions often throughout the course, using activities and lab exercises—all ways for you, the instructor, to visually or audibly "listen" to your students.

Adult learners benefit immensely by hearing what their fellow students are learning. Don't limit the conversation of learning to one between you and your students, but encourage the conversation between your students as well. When one student asks a question, allowing another to answer creates an opportunity for this type of learning. Adult learners learn best when they hear the same information from multiple sources and a classmate is another source each student can learn from.

The Foundation for Engaging Learning

I have already discussed and referred back to the idea that adults learn by doing. Engagement is a way of doing and, as such, is a great learning tool, but there are other reasons why engagement assists in the learning process as well.

Students Learn Better When They're Awake

Okay, so you didn't buy this book to learn the obvious. That's the point. It's obvious. You do not need to be told that students learn better when they are engaged. I could have stated that students learn better when they're having fun, though for some, depending on your product, that might be a challenge!

Many people believe that the primary reason for engaging students is to keep them from falling asleep. They may not state it in those words, but that is why many presentation instructors encourage it.

Learners Require Time to Absorb the Learning

The real reason why adults need to be engaged is that learning takes time. Engaging your students provides time for information to soak in. One common rule of thumb used by many experts is that learners typically require 2 minutes to absorb every 10 minutes of material. Absorption happens best when the brain is allowed to shift, even slightly. When another student asks a question, when the instructor asks a question, when time, movement, or physical action is required—all of those contribute toward the 2 minutes of time to soak in.

Breaks to use the restroom and refill your coffee are important, but I am not including those in the 2 minutes of soaking in time. You should be asking questions and getting student involvement regularly. Never go more than 10 minutes without some sort of deliberate engagement of your students.

Set the Expectation for Engagement

Throughout this chapter I will encourage you to engage your students in a variety of ways. You must set the expectation to be engaged early. If you put off engaging your students until you have first delivered several hours of lecture, don't be surprised when they aren't enthusiastic.

Tell your students that you expect them to take ownership of their learning by engaging in the activities, asking questions, and volunteering for assignments. They are very aware of personality differences. Encourage those that tend to be more aggressive to get others involved and those who lean toward simply watching to step out of their comfort zone. If you don't set the expectation at the beginning, you forfeit the right to make that happen if you need to later on.

Practical Engagement in the Classroom

It is one thing to believe that engagement is important, to really grasp that learners learn best while they are doing something, instead of just demonstrating their learning after the fact, and that they need engagement to help them absorb the content. It is quite another to put that into practice in the classroom. It takes practice, trial and error, and a tenacity to stick with it, even when things don't work as you'd like. I'm going to get you started with some ideas, but the ideas are limitless and vary significantly depending on your product, your objectives, and your environment. The best way to learn what works for you is—by now you've guessed it—by doing.

Engaging as a Conversation

The most practical thing you can do to switch from presenting to facilitating is to turn your training into a conversation. The more respect you show your students, the more respect they will return to you. Remember all adults bring something to the conversation, even if they are brand new to your product. Sometimes, those that are new will provide tremendous insight and ideas that haven't been biased by their familiarity with your product. Learn to use that to help everyone in the class learn. Here are some practical ways you can do that. First, however, in the spirit of this same idea, take a moment and write your own ideas for doing that in the box below. Even in this conversation, you have ideas you have used or seen used. Feel free to send them to me as well. I'm always trying to learn how to be a better communicator.

> One way I can teach my students while involving them in a discussion.

I'm sure your way is great. Because you know your circumstances and products, it is probably much better than anything I will suggest, but I'm going to make three simple suggestions anyway:

1) **Let them have a say in the order things are taught**

   Okay, you might be panicking right now. It is true it takes some practice, but I'm not talking about changing the outline around. Very often, in product training, I find that there are bullet points or lists of various topics that need to be covered with the students. Instead of showing one bullet at a time, as it magically flies on to the screen, you might simply list the bullet points and ask the students which one they want to cover. I find that the simple act of going "out of order" keeps students engaged and active. I might go down the list, calling on the next person to say which bullet point we are going to discuss next.

   Another way to do this is to put the points to discuss on notecards. Randomly distribute the cards and ask different students to read their card, stating what the next topic will cover.

   This type of activity accomplishes several things. First, it ensures that all students have an equal voice. If you leave participation up to the most vocal student, quieter students will not be able to contribute. Second, this activity forces students to stay in the conversation, even if their input is as simple as reading the next topic point. Third, your use of creative engagement demonstrates a level of preparation and thoughtfulness that many are not used to in technical training. Teaching with interaction takes practice, but it has tremendous rewards that are observable and lasting.

2) **Use their experiences and backgrounds, not just your own**

   Your own experiences are a great teaching tool. They are what make you unique as an instructor. Use them. Students expect that and it can help to increase their willingness to learn from you. But the irony is that the most experienced experts are the ones most willing to listen to other people's experiences. You shouldn't do so with the hopes of impressing them with a better experience, though! Make sure it is genuine.

   Using your students' experiences can be challenging from a time management perspective. One possibility is to turn the idea listed above around and do the exact opposite. Hand out notecards with the topics or bullets to be discussed. This time, however, you may want to allow students to choose which card they pick. As you get to that bullet point, ask the students why this point is important, or for a short example of why it matters. If you keep things moving, it can be an engaging and profitable discussion, and you will find that you actually did very little teaching.

   Whatever you do, don't fall into the habit of re-teaching everything. If your students give a good enough reason, swallow the ego, state that it was a great reason, and move on.

3) **Ask for volunteers to help**

   Next time you do a demo, don't do it. Let someone else perform the demonstration. Unless, of course, the demo is a marketing trick to show how easy it is for an expert with multiple years of experience to perform the task. Your goal is for your students to learn, not to showcase your talent. The main talent you should be showcasing is your ability to teach. Guide and coach your students as they use your equipment or design a solution. Will this approach slow you down? Absolutely. But the investment in time will pay dividends in learning.

I once helped an expert who was seeking to demonstrate a software product to a class. After watching him demonstrate the product, moving around the screen at lightning speed, I was lost, as were most of the students, who rightly figured that they would have to learn it for themselves later. I was reminded that we are very accustomed to technical training really being a demonstration about what we will learn on the job later.

I encouraged the expert to have a student navigate the software the next time. He assured me he understood that it would have been better learning if everyone had a copy of the software (he was right), but that since they didn't have enough copies, he felt he should just demonstrate the product to them. He agreed to change his thinking, and the results were phenomenal. The attention of the students when another student was at the controls was completely different. The student was slower, so the rest of the class could follow along, and the entire class "helped" the volunteer find the right places to click, and so on. The learning experience was vastly different, simply by putting a student at the controls.

Engaging with Questions and Answers

Questions are your best means of engaging students. Questions can be asked in two different directions. Students can ask questions to you, and you can ask questions to the students.

Why Instructors Ask Questions
There are several reasons why instructors ask questions. Here is a simple list of why you might ask questions, as an instructor. Feel free to add to the list your own reasons:

- Check for comprehension or understanding.
- Get the class involved or focused.
- Give time to absorb, or break up longer ideas.
- Reinforce key points (even questions with obvious answers help to do this).
- Start a discussion or conversation.
- To get answers! Yes, it is okay to ask questions because you really need an answer.

All of those are valid reasons why you might ask your students questions. Whatever your reason for asking the question, do not forget the most important thing to do after you've asked one.

Wait for a response!

That's right. Many new instructors are so afraid of silence that they are very quick to answer their own questions. When you do that, you send the message to your students that you do not really expect a response and you may not get one when you do. Your students need time to process the question and formulate an answer. If you don't give them that time, you are highlighting the difference between your knowledge and theirs, something students perceive as disrespectful. The reality is that you would probably require the same amount of time if you had to process both the incoming information and the outgoing response as well. Don't be afraid of the silence.

When and How to Ask Questions

Begin asking questions as soon as possible. I often ask a question that requires an answer from everyone as the very first thing I do in a training class. There is a purpose for this. Asking questions early in the training session will set the tone for the rest of the training. The expectation that they will need to be engaged and ready to participate in the conversation of learning is highlighted. They cannot sit back and halfheartedly listen and check a few emails on the sly.

Start the session with easy, nonthreatening questions. Ask simple closed-ended questions and unobtrusive open-ended questions. After they have become more comfortable with each other, with you, and with the content, you can start to ask more difficult questions. If participants are encouraged to answer questions at the beginning of a class, they will be more likely to continue answering questions as the class proceeds.

While asking the right questions is a great way to engage your students, asking the wrong ones is a sure way to alienate them. There are two types of questions you should avoid. First, avoid asking vague questions. Many instructors ask questions that no one can really answer. For example, "Does everyone understand that?" is not a question anyone can answer, unless they have some insight into everyone else's understanding. Instead, ask specific questions, even if it is directed at the entire group. "If someone asks you, will you be able to explain this concept?" is better than the generic, "Does everyone understand this?" Second, avoid questions that are intimidating. Instead, ask questions that encourage learning, not ones that embarrass them or make them feel uncomfortable.

Figure 12.1 shows examples of vague and intimidating questions on a grid. Practice using questions that are closer to the top right of this grid. Avoid the left side altogether.

Figure 12.1 Diagram of questions.

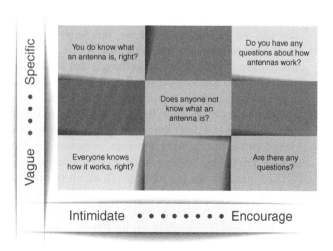

Answering Student Questions

Just as you should ask questions early and often, you should encourage your students to ask questions from the beginning of class. With the possible exception of a webinar that requires muting of phone lines, do not include a "Q & A" slide at the end of your training session. If your training is a conversation, questions should be asked throughout the event. Telling students to hold their questions until the end of class indicates a stronger desire to get through the material than to generate learning. That may be appropriate for a presentation, but it is not appropriate for facilitating proficiency.

Allowing questions throughout your conversation has its challenges. You have to remember that it is still your responsibility to meet the objective of the class, and sometimes you won't be able to answer all of the questions. If questions are asked that are going to be covered later in the curriculum, write them down and let the student know it will be covered later. Use a flip chart as a "parking lot" for unanswered questions, or otherwise visibly note the questions. This assures students that you aren't just going to forget about their question.

There are four general principles I try to follow when answering questions from my students. None of them are hard and fast rules, and almost all of them will need to be tweaked or even broken from time to time, but adhering to them in principle will gain you respect and appreciation from your students:

1) **Restate the question**

    Restating a student's question does three things. First, it allows everyone to hear the question. Second, it verifies to the student that you understood the question. Third, it gives you that much more time to collect your thoughts and answer appropriately.

    This is also true of questions you get during a break or outside of the classroom. Ask the student if you can share it with the group, and then do so. Usually, if one student asks a question, more than one needs the answer.

2) **Give others an opportunity to answer**

    This cannot be done all the time, but significantly more than what is typical. Allowing other classmates the opportunity to answer the question keeps them engaged and allows students to hear the same thing from another perspective. This practice can be challenging, but is rewarding. Make sure the opportunities are spread around the class and the question is answered completely and thoroughly.

3) **Don't be afraid to say, "I don't know"**

    Admitting you don't' know something is not a blemish on your character or a flaw in your expertise. The reason you have been asked to be an instructor is because you know more than most. If you don't know it, that is fine. Only those who know very little think they know everything. Make a commitment to your students that if they ask something you don't know, you won't make up an answer.

    Students almost always show an increased level of respect after an instructor says, "I don't know." Some students just need to hear that. They ask increasingly difficult questions to see when you'll admit to not knowing something. What they really want to know is that you will be able to coach them toward more learning, even if you don't know everything. My advice? Get it out of the way as quickly as possible. Once you have, you'll be free to move on. Of course, if you don't know something, be sure to get back to your students with the answer.

4) **Get their permission to move on**

Nothing will discourage your students from asking further questions than the feeling that it is useless to do so. After you answer a question, find a way to get their permission to move on. You can ask, "Did I answer your question?" if you feel that it did. If you were unable to answer the question, either because you didn't know the answer or because you are going to cover it later, just ask them if it is okay to move on. If you don't do that, you risk losing them at that spot, as they try to sort through why you didn't answer their question.

As you can see, asking good questions takes practice. When you do it well, you will find that questions are a great way to listen to your students and engage them in a mission to discover more learning.

Engaging Group Learning Activities

Another way to engage your classes is to use group learning activities. Brainstorming sessions, group discussions, or anything that uses the collective knowledge of the group instead of just the singular knowledge of the instructor is a great way for adults to learn.

Group activities take considerably more thought than questions and answers in order to effectively engage your students. You should test any activity on a pilot group. There are some specific challenges related to group activities, but each has solutions that are not difficult and worth the extra effort:

1) **The challenge of clarity**

There are two main challenges for the student, and they need to be clear before you break them into groups. The first involves what they are supposed to learn, and the second involves what they are supposed to deliver.

- *Clarity of the learning objective.* Make sure the learning objective is clear. Clarity of objective is always important, but when you are up front teaching, students will tend to be more forgiving about it, assuming you will make that clear eventually. When you break students into groups, they intuitively realize that you can't be with all the groups at once. They immediately go into the "we are teaching ourselves" mode, which, to be fair, is exactly what is happening! Be sure to give them what they are supposed to discover or learn about in some sort of finite time or content boundary.
- *Clarity of desired deliverable.* Clearly define how you want them to capture what they are learning. It is helpful to write down the instructions to pass out. Or you could distribute a worksheet. In any case, the learning process must be clearly achievable.

2) **The challenge of time management**

Always give clear parameters regarding time and stick to it. Put the time on the worksheet and set a timer, if that helps to keep the class on schedule. This includes the review at the end. If you're not careful, some students will take 15 minutes to review their 10-minute discussion. Make sure you set the parameter to discussing only what was discussed in the group, not their hobby horse. Keep it moving. You should almost always feel a little pressed for time. You want to keep the groups active and avoid having a lag time that risks losing their attention.

3) **The challenge of total involvement**

Keeping everyone involved is the whole point of the group activity, so if students begin to disengage, it is especially bad. Methodically review the findings of each group with the whole group. Doing this will diminish the possibility that one group learned something another one didn't.

Choose a spokesperson to represent the group. If you keep the group for a couple of activities, be sure to choose a different spokesperson each time they present.

Engaging Labs and Exercises

Labs and exercises are a crucial part of product training. It is often the labs that will dictate the maximum number of students you can teach. The labs will elevate your training from a boring session to engaging, hands-on proficiency instruction.

The labs are not the only time you should engage your students, but they are important. This is an opportunity to move beyond the first stage of hands-on learning and allow your students to explore your product. As a reminder, Table 12.1 lists the three levels of hands-on learning that were introduced earlier.

Divide the class into small subsets so that each student has an equal opportunity to perform the required exercise or to explore the product. The learning principle here is that what you allow the students to discover on their own, they will retain much longer. They will also have a sense of satisfaction that they completed a task, which adds to the class and keeps them engaged.

Icebreakers, Games, and Other Interactive Options

We are fortunate to have so many wonderful options for interacting with our students. Icebreakers, games, and other engagement options play an important role in a facilitated training environment, even when the subject is technical. Icebreakers are not just for soft skills training, and they are not just for "breaking the ice." Icebreakers in proficiency training can include many types of student involvement.

Icebreakers can be called any number of things. Introductions, opener, discussion starter—all are other names for icebreaker. What you call them is irrelevant. That you do them is important. Most often, the best thing to call it is—nothing. There is no need to give it a name or even tell your students that you are doing one.

Why are they helpful?

Table 12.1 Stages of hands-on learning.

Approach	Driver	When to introduce	Retention	Response
Exhibit	Product	As early as possible	Short	"I understand it"
Execute	Instructor	After sufficient introduction	Medium	"I can do it"
Explore	Learner	Throughout	Longer	"I did it!"

Susan D. Landay lists 14 reasons why trainers use icebreakers[1]:

1) Allow participants to express their expectations
2) Introduce participants to one another
3) Build a sense of community
4) Introduce the content
5) Set the tone for the session
6) Help get conversations going
7) Help people remember names
8) Get people on their feet and get the blood flowing
9) Engage participants in the learning process and set the tone for participation
10) Give participants a sense of ownership over the learning
11) Break down barriers between the trainer and the participants
12) Encourage participants to think differently
13) Understand the knowledge and experience of participants
14) Enable participants to network with each other so they can use one another as a resource after the training has ended

Adults learn better from multiple sources. In the classroom, that other source is likely a fellow classmate. To cultivate enough of a relationship to learn from each other, students must first get to know each other.

When Should They Be Done?

Icebreakers are not only for the beginning of class. They should also be done throughout the class. You can do an icebreaker to review a lesson or simply because you sense the need to re-energize the students. Do them for any reason, as long as you have a reason. Don't ever do them just to do them. Make sure your goal is accomplished and that you make the connection between the event and the intended outcome for the students.

Games and Gamification

Games can be a great way to stimulate learning, even with adults. Early in my career, I felt games were for training departments that didn't have serious information to learn. At one point, I was struggling to convey a networking concept to non-networking professionals, and it was proving to be difficult. Eventually, I made a game in which the participants tossed a tennis ball around the room, simulating a network packet being bounced from one location to the other. As the students played the game, they helped each other determine where the ball should go next. As they did, their eyes would light up and I could see them learning the concept. I became a believer in games, and in what the training industry refers to as gamification.

Games can be used in a similar way to icebreakers, though they often include some sort of competition, either as individuals or as a group. There are many resources for gamification on the Internet, though most of them are geared toward enterprise learning, not product training. The main thing you want to consider when creating a game is the purpose. Games are often used to extend training beyond the classroom or to tie multiple training events or resources into a cohesive and competitive whole. Any game you create should be for a specific learning purpose. If it is played in class, it should be quick and effective. When done well, you can use a game as a fun and memorable way to convey a concept.

Interactive Technology

Another way to effectively engage your students is to invest in technology that allows them to participate in the learning with you. I use a polling tool in all of my classes. I find that it forces me to prepare questions for the class and it keeps them engaged. It also has a benefit that I had not considered before I started using the tool. It helps me to slow my teaching down to the rate of absorption. If you just ask a question to your students and one person answers it, what do you generally do? You move on. But if you ask the question using an interactive process, you can see if all the students are ready to move on or not. This small detail has been very helpful to increase the effectiveness of training.

Conclusion

When you stop talking, your students will absorb more learning. When you listen to your students, they will listen more intently to you. When you are engaged with them, they will be engaged with you. When you shut up and listen to your students, you become a better instructor. Find creative ways to get them to discover the learning for themselves without your voice having to be the only medium to relay the information. Ask questions and then get out of the way. Break them into groups and then let them teach themselves. Pair them up and let them explore on their own. Give them exercises and let them complete them alone or in groups. Be creative. Stretch yourself as an instructor. Find new ways to teach the same thing, but make sure you engage, engage, engage.

But student engagement is about so much more than the time they are in the classroom. Students that you engage in the classroom will be excited about taking that learning beyond the learning event itself. Isn't that what you want? There is no possible way you can take a group of people and turn them all into experts in one class. But you can send them out to become experts...

...If you got their attention. If they were engaged.

Making It Practical

Teaching is a two-way conversation. In order to engage our students, we have to allow them the opportunity to speak and do. We listen, not only with our ears but also with our eyes.

1) List three ways you can "shut up" when you are teaching your next class.

2) What is your biggest challenge to creating an engaging classroom environment? How have you or will you overcome it?

Before you read Chapter 13, "Stand Up: Effective Nonverbal Engagement," answer these two questions.

1) What do you believe has the largest impact on your credibility as a public speaker: your verbal communication skills or your nonverbal communication skills?

2) Describe how the layout of a classroom might change your teaching approach.

Note

1 Landay, Susan D. Not Another Icebreaker! http://elearnmag.acm.org/featured. cfm?aid=1966301 (accessed August 8, 2017).

13

Stand Up

Effective Nonverbal Engagement

Your students can completely trust you and believe what you are going to say even before you open your mouth! Now that is encouraging! In a well-known (and often misrepresented) study completed by Dr Albert Mehrabian[1] at the University of California, only a small percentage of your credibility has to do with the actual words you speak. Words are important, but what will make others believe and trust you has more to do with how you say those words and what you look like or do when you say them.

While research may help to put a percentage number on the statistic, anyone with teenagers—or, for that matter, ever was a teenager themselves—knows this to be true. I've heard (and likely said) the words, "Yes, Dad" in a variety of different ways, each with a significantly different meaning. The point is that the way we say things matters, even in the classroom.

How smart or educated you are doesn't matter; you don't get a pass on nonverbal communication. All instructors must learn to engage students nonverbally. All communication must be encoded by the communicator and then decoded by the receiver. As you read this chapter, think of at least one thing you can do to improve your observable communication and one that you can do to improve your perceived communication. Put yourself in the place of your students. They are listening to you, hearing you say things about your products. Does what they see you say align with what they hear you say? While they are listening to you and observing you, they are also perceiving things about you.

Be sure that what your students see and hear correspond with what you are actually saying. When they do, you will capture your students' attention. Knowing that what students see influences how much they believe will encourage technical instructors. What an easy way to influence your students' perceptions about you without even changing what you say!

Observed Communication: What They See You Saying

If what your students observe when you tell them about your product is at least as important as your actual words, it would be irresponsible not to ensure that they see the right thing. There are at least four common ways we communicate with our bodies

Product Training for the Technical Expert: The Art of Developing and Delivering Hands-On Learning, First Edition. Daniel W. Bixby.
© 2018 John Wiley & Sons Ltd. Published 2018 by John Wiley & Sons Ltd.
Companion website: www.wiley.com/go/Bixby

in any live setting. There is a fifth way, an instructor's proximity to his or her students, that may be somewhat unique to a facilitation environment.

Posture

Observed communication is important because it influences perceived communication. One great example of that is posture. Posture is the way you hold your body. Posture communicates several things, but one that is important to you is confidence. You can look more confident by changing your posture. In fact, it will not only affect your students, but it will affect you as well!

See if you can match the six different postures on the left with the interpreted message on the right. The point is not get them all right as much as it is to think about the fact that our posture communicates something to our students. The answers are at the end of the chapter.

Exercise 13.1 Posture exercise

It is generally true that the more confident humans get, the more space they are willing to occupy. When you are afraid you want to get small, so you hold your elbows to your sides or fold your hands in front of you. The more confident you become, the more space you feel comfortable filling around you. If you struggle with confidence, put your feet shoulder width apart, spread your arms, make yourself big, and conquer the world! You will be amazed at how much better you feel just by changing your posture, and your students will feel the same way.

As you observed in the exercise above, posture does more than communicate confidence and control. It could communicate fatigue, disinterest, sincerity or lack of it, comfort or nervousness, and probably a whole lot of other things. You will also note that it doesn't always mean the same thing to everyone. For example, I'm quite sure that some of you matched *one hand in pocket* with "He's one of us" when I would have matched it with "He's not taking us seriously." Does that mean that you're wrong, and

I'm right? No, but it does tell us to be careful about putting our hands—or even one hand—in a pocket, because someone might misinterpret it. The goal in all of these is to eliminate any interference in the communication. Taking your hand out of your pocket just reduces the chances and it's a small price to pay.

Facial Expressions

Almost every day as I leave for work or get ready to teach a class, I get a friendly reminder from my wife. "Don't forget to smile." Apparently, I sometimes forget. I hope, of course, that I forget less often now, as a result of her faithful prodding. Facial expressions are important and they can be practiced. If you aren't lucky enough to have someone who cares enough to give you good feedback, go stand in front of the mirror or record yourself. Make sure your face is saying the same thing as your words.

Smiling is not, of course, the only facial expression. The key is to be genuine. Facial expressions are especially important when we are listening, which is part of communicating. Have you ever told someone something and been able to tell by their unchanged expression that they weren't listening to what you were telling them? The same is true with your students. Make sure they know you are genuinely interested in communicating with them and show it on your face.

Eye Contact

Even more important in the conveyance of sincerity and honesty is an ability to make eye contact with your students. Don't use your screen or any other prop as an excuse for not looking directly at your students. This is especially important in a proficiency training session. Remember you are teaching individuals and they need to be contacted visually.

Here are a few tips I try to follow regarding eye contact:

1) Make eye contact with everyone. Don't single out one or two "friendlies" and forget about the rest of the class and don't scan across their faces. Pause when you make eye contact with someone.
2) Figure out a comfortable time. This depends a little on culture, but all cultures have a happy medium between shifty eyes and awkward. It's usually a few seconds long.
3) Randomize your eye contact. Any nonverbal communication can look staged and uncomfortable, including eye contact. Going from one person to the next down a systematic line only aggravates that potential. Randomly looking at people will keep them more engaged and help you look more natural.
4) Face them, don't just look at them. When you face someone you pull them into a conversation with you, instead of making them feel like they are being talked at.

Eye contact, of course, changes in certain circumstances, like when you are answering a question (more direct) or writing on the whiteboard (less direct). It also may vary by culture, especially in cultures where eye contact between the sexes could be considered offensive. The general principle behind eye contact, however, is that you should be communicating to an individual student, moving quickly (every few seconds) to another individual student, so that all feel like they are involved, but that they are involved as individual contributors and receivers of the communication.

Gestures

Gestures may be the most culturally flexible of all the ways we communicate nonverbally. As I have traveled and taught around the world, I find we have so much in common in regard to what people understand. There seems to be a distinction between formal gestures, which are often used for greeting someone or recognizing their presence, and informal gestures in conversation. This is highlighted in Terri Morrison and Wayne Conaway's best-selling book *Kiss, Bow, or Shake Hands*.

After that initial and formal greeting, gestures seem to normalize. Mine, of course, is not a formal study of cultures, but I have lived for over a decade on three different continents and trained all over the world as well. It is, of course, a good idea to study something about any culture you are visiting, or rely on a local colleague so that you don't offend anyone or unnecessarily take offense at an innocent gesture.

Here are a few tips about gestures:

1) Figure out what to do with those darn hands.
   a) Do NOT put them in your pockets.
   b) Don't hide them behind a lectern.
   c) Don't point. I prefer an open and upward-facing palm, though some use a thumb and closed fist to point.
   d) Don't play with something. Holding something is fine, as long as it doesn't become the source of a repetitive or annoying gesture. You don't need a pen or a remote presenter to keep you busy if what is being communicated is important.
2) Use gestures (and posture) to speak to yourself, not just your audience.
   If you aren't 1 of the 33 million plus who have already watched Amy Cuddy's TED Talk on body language, I would encourage you to do so. It is a powerful reminder of how our body language speaks, not only to our listeners, but to ourselves. Even if you don't struggle with nerves, I highly recommend taking a few minutes to watch it.
3) Practice, practice, practice.
   The next time you watch a symphony conductor wave his or her arms in front of the orchestra, ask yourself how much practice that took. I can guarantee you they spent hours in front of a mirror or professor learning precision movements and fine-tuning their skill. The same is true of your simplest learning gestures. If you are used to keeping your hands in your pockets, it will seem awkward to have them out in the open. With practice, it will become natural. Soon it will look like you've always done it that way.
   If you have a local speaking club in your area, I would encourage you to join it. The practice along with the feedback is a valuable way to continue improving.

Physical Presence

There are many ways to talk about where to stand or position yourself when you are teaching. All of the other nonverbal communication areas I've mentioned will change very little as you transition from presenting to facilitating. Your physical presence, however, may change significantly. For many lectures or presentations, the speaker is on a stage or behind a lectern. Now, you likely won't even have one. When you realize you are going from being a professor to being a coach, your physical presence—the way you move around the classroom—will change.

Here are a few tips for creating a good learning presence:

1) **Move. Move deliberately, but move**

   Don't teach from one location. As much as the environment allows, move around. When you do move, move with a purpose. Don't pace or walk around aimlessly. If the room has natural divisions, move to one side, stay there for a short while, and move to the other side.

2) **Give equal proximity to everyone**

   As much as possible, get equally close to everyone while you're teaching. Worry less about presentation techniques, like which side of the slides to stand on, and more about engaging individuals in a conversation. To do that, you will need to face them and move toward them.

3) **Don't talk to the screen**

   Engaging facilitators don't have to be great orators, but they do need to talk to their students. Talking to the screen instead of to your students indicates a lack of confidence in the material and nervousness. If you are reading from the screen, it likely indicates that you are putting too much text on it. If you only put pictures or very short reminders on the screen, it will be harder to read it.

Physical Appearance

There are no rules about how trainers should dress or what hairstyle will best aid in a transfer of knowledge. What we do know is that appearance can influence people. Appropriate attire will change depending on your audience and your training topic or location. It must be appropriate. Never assume that the salesperson has to worry about attire, but the technician does not. If you are training on a product, you influence sales, much like a mechanic might influence your opinion of a car as much or more than a salesperson. When you stand in front of a class to represent a company, you ARE the company to your students. Dress like it.

All of these areas are important to consider when you are facilitating proficiency training. Take the time to consider one thing you are going to do differently in your next training class. Write down one area you want to improve, along with a specific action you are going to do differently. Come back after the class and take the time to reflect on what worked and what did not.

Exercise 13.2   Observed communication personal improvement goal

Area I want to improve:
Specific thing I am going to try:
Reflections. What worked and what didn't?

Perceived Communication: What They Feel You Are Saying

Not all communication is observable. Some is just felt. This isn't difficult psychology; this is just common sense. Of course, all of the observable areas I just covered strongly influence the non-observable areas of communication, but it is helpful to state at least a few of them, as they pertain to the technical training classroom.

Be Genuine and Humble

Adult learners want a genuine instructor, not a perfect one. You should genuinely want your students to succeed. If you don't, stop teaching until you have a change of heart.

Students are very adept at deciphering the difference between a genuine desire for their success and a genuine desire to impress them. They can tell if you have a real desire for them to learn or just really want to get the class over with. They are not always right. But whether they are right or wrong doesn't change the fact that it will influence how they decode the communication you are sending to them.

You cannot be genuine without being humble and honest. If you don't know it all, don't pretend to know it all. You don't need to be a guru to be an instructor; you just have to have enough guts to lead people to good learning.

Be Likeable and Pleasant

Students are more eager to learn when they are enjoying it. That doesn't mean that every learning module must be a game or a comedy show. It does mean that you can keep them engaged much more easily if they like you and you relate well.

Don't focus on just one or two students. Treat everyone with equal professionalism and kindness. Even though students are out of their normal work environment, it is still important to maintain a professional and pleasant environment. Avoid crude language or anything that might be offensive to any of your students, since those things may cause a distraction from their learning.

Be Available and Prepared

Training starts the instant the first student walks through the door. If you are busy with last-minute preparations for your class, you will not be available for your students. Arrive to training in enough time to get fully ready to start before any of your students arrive. For me, that is usually an hour prior to training. Check your equipment, go over your materials, and make sure you are completely ready before your first student arrives.

If the class size permits it, greet every trainee with a warm smile and handshake before beginning the class. Doing so sets the tone from the very beginning that you are interested in teaching them as an individual, not just in delivering a message to a group of people.

Be Positive and Have Fun

Adults are not much different than kids when it comes to their desire to enjoy learning. Humor, when used appropriately, can lighten otherwise dry or intense subjects. But having fun doesn't have to be extremely overt. Students can tell if you enjoy what you are doing or not. Instructors who get frustrated easily or are negative with their students are communicating that they would rather be somewhere else. Students will get wrong answers. They will do foolish things—that is why they are there! They need to learn. Be excited about seeing progress; don't focus on where you wish they could be so your job would be unnecessary.

Of course, do not use humor that might be offensive or distasteful. The point is to enjoy your time together. Don't get so focused on the task of training that you forget to get to know your audience. Just as your students will not be more excited to learn than you are to teach, they will not enjoy learning from you any more than you enjoy teaching them.

Be Confident and in Control

It is very possible to be both confident and humble. Those are not opposite ends of the same continuum. Your students expect you to be both, and they will learn best when you are. Staying in control of the classroom is not about ego, but it is about an honest desire for everyone to learn. It takes confidence to stay in control. Whenever I've seen a classroom get off track or an instructor begin to lose control, it has been because they lost confidence in their ability to lead the class.

Use everything you have read in this chapter about nonverbal communication to demonstrate confidence and stay in control. It's harder to lose control of students you are looking in the eye, smiling at, and being friendly to. When students like you, they will want to help you do your job well, and proper preparation is one of the better ways to ensure confidence before the class even starts.

One of the more useful and subtle nonverbal techniques to maintain control is physical proximity. When you are regularly giving equal proximity time to all of your students, you will rarely have to deal with talkers or email checkers. If you do, you can simply move into "their space" and they will quickly adjust their reading habits. If they don't, one helpful trick I learned early in my career is to put my hand on their table or desktop. I don't address them. In fact, I rarely look at them. However, just the simple act of touching their personal space gets their attention. If they continue to check email at that point, I know that something serious must be going on and I generally leave it alone until a break.

Verbal, observable nonverbal, and perceived nonverbal communication all jive together like an intimate dance. Overemphasizing one and neglecting another is conceivable. For that reason, I encourage you to choose one thing you are going to do differently in your next training class. Just as you did for the observable communication, write down one area of perceived communication that you want to improve, along with a specific action you are going to do differently. Come back after the class and take the time to reflect on what worked and what did not.

Exercise 13.3   Perceived communication personal improvement goal

Area I want to improve:
Specific thing I am going to try:

Reflections. What worked and what didn't?

Environmental Influences

Great communicators can excel in the harshest of environments. The rest of us shouldn't take that chance. The environment you give your training in influences how your students will decode the information you are encoding. If you can make the learning easier, you should.

If you have the opportunity to create your own learning environment, make sure to get a lot of advice. That is an opportunity you want to take seriously. Things like lighting, room dimensions, floor outlets, and so on, are all pieces to a bigger puzzle, but a puzzle that adds up to better learning. I recently read an email from someone who was complimenting one of my trainers on having the best training they'd been to in their 30 years of working in the industry. The compliment eventually mentioned the instructor's skills, but first, it brought attention to the training facility and how it aided in the learning. Environment matters.

Room Layout

If you don't have control of which room and how bright or controllable the lights are, you might be able to move the tables around. I like to set my product training rooms up in a U shape. The U-shaped room has several advantages:

1) There are a limited number of seats available.
2) All students are front-row students.
3) The instructor can be equally close to all students.

All three reasons are important. Students cannot hide in the back row when there isn't a back row. The U-shaped classroom immediately sends the message that they will have to be engaged during the class. From the instructor's perspective, it is easier to control the classroom and engage the students in this setting.

Furniture, Lighting, and Technology

Chairs should be comfortable enough to sit in for the duration of your class. If you are able to, get chairs with wheels on them. This encourages students to pair up, move into groups, and so on. Remember you won't be sitting still for long at all. Chairs with wheels help to convey that concept from the beginning.

Use technology that gives you freedom to teach from anywhere in the classroom.

Lighting is tricky if the room wasn't designed as a learning environment. If you have no control of the lights, make sure your font sizes are large enough to read from the back of the room in the lighting you have. Good background color and contrasting text color are also helpful. You may also want to give more breaks, if the lighting seems to create fatigue.

Know Your Environment

It is one thing to train in a room you helped to build or are very familiar with. It is another to show up at a customer site and be completely surprised. I once trained in a

Figure 13.1 U-shaped classroom.

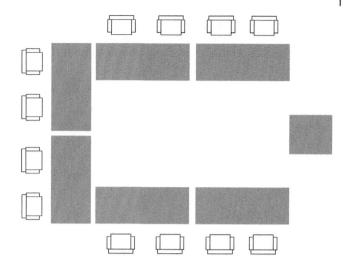

very small communications room directly off the tarmac of Atlanta International Airport. To put it mildly, it was not a great training environment. Ever since, I've made sure to avoid surprises.

Send a drawing of how you want the tables set up, including how large the room should be and how many students should be seated at each table to your customer or hotel conference room in advance. Don't expect them to know what a good training environment looks like. It doesn't have to be complicated. Something simple will work. If you can, ask for pictures of the room ahead of time. Sometimes you can't control the room, but knowing what it will be like ahead of time can help you prepare (Figure 13.1).

Hosting a Training Event

As a technical trainer, you are more than just an instructor. You are a host. Even if you are stepping into an unfamiliar room, if you are the teacher, it is your responsibility to make sure your students are comfortable and ready to learn.

Make Your Students Feel Welcome

From the minute your students step into the classroom, it becomes your job to make sure they feel appreciated and welcomed. Making them feel valued as learners goes beyond the handshake I suggested earlier. That is just the beginning. You need to make sure you do everything in your power to make the learning easy.

Making students feel welcome isn't accomplished through some sort of mystical means. Make them feel welcome in tangible ways. Here are a few things to think about, but add to the list as needed for your situation.

Prior to arrival
- Send directions to the location and room.
- Ensure they will have access to the building.

❑ Communicate any prerequisites or requirements.

❑ Ask about any food allergies or restrictions.

❑ Provide a local contact name and phone number to call if they have problems.

Upon arrival

❑ Use signs or people to provide direction.

❑ Greet each student with a warm handshake.

❑ Assist students in finding a seat and meeting other students.

❑ Be prepared for questions regarding electrical power for their electronic devices.

❑ Provide the Wi-Fi access code, if necessary or appropriate.

❑ Ask about their hotel stays and show genuine concern for their comfort outside the classroom.

At the start of the class

❑ Inform students of safety procedures, evacuation routes, building procedures, smoking areas, restrooms, break rooms, and so on.

❑ Provide name tents or name tags and markers (unless preprinted).

❑ Cover any ground rules or policies to maintain professionalism and eliminate offensive behavior.

❑ Provide writing utensils and paper at minimum. If possible, work with your marketing department to provide other giveaways.

Throughout the class

❑ Provide plenty of water to keep students hydrated.

❑ Provide coffee and snacks.

❑ Provide lunch, or at least assist students to find nearby lunch venues.

❑ Consider having seasonal items available to borrow: umbrellas, winter jackets, sun screen lotion, and so on.

After the class

❑ Assist with any travel concerns or questions.

❑ Provide suggestions for local eateries and entertainment.

❑ Thank each student individually and encourage them to continue learning.

The goal of all of these items is similar to when you have guests in your home. You want them to feel valued, comfortable, and appreciated. Doing so is, of course, the polite and right thing to do. But these simple acts of hospitality also have a learning benefit. They help to eliminate any obstacles that may otherwise prevent a student from learning. No one wants to spend hours preparing content and fail as an instructor because a student couldn't find the company cafeteria. Caring about the little things outside of the classroom will make the learning inside the classroom even better.

Conclusion

What we say as instructors has less impact on our students than how we say it. How they remember us as an instructor is going to be more affected by what students perceive about us than what we actually say. Their attitude toward learning affects their learning. Your attitude about teaching affects their attitude toward learning. Make sure that your love for technology translates into a love for the people learning about that technology. Teachers are nothing without students. Treat your students with respect.

Having guests is hard work. You have to clean the house a little bit better, make a little bit more food, and think a little bit harder about the things you take for granted. The same is true with training. If you don't know your training environment, do your best to get as much information as possible about it.

The best way to know your environment is to see it yourself. Get there early enough to make adjustments if you need to. More importantly, get there early enough to test your equipment. If they are providing the technology, make sure you know how to use it. If you are using a sound system, test it. Be prepared to host your students and serve them for the duration of the training. Test everything and get ready for a great class!

Answer to Exercise 13.1 (Figure 13.2).

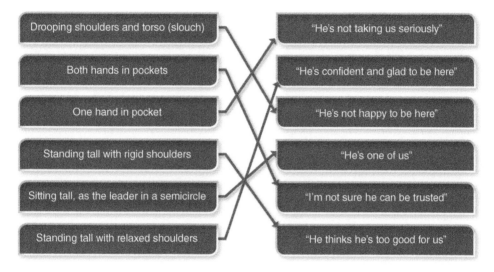

Figure 13.2 Answers to Exercise 13.1.

Making It Practical

Your body language can set you up for success, or it can betray you, in spite of your best intentions or greatest content. Considering how others are receiving your message is important.

1) What is one area of nonverbal communication that you want to change in your next class, and how are you going to change it?

2) How can you use nonverbal communication methods to stay in control of a classroom?

Before you read Chapter 14, "The Smartest Engineer: And Other Difficult Students," answer this question.

1) Think about the most challenging student you've ever had in a class. How did you handle him/her, and what would you do differently if you could do it again? (If you are a new instructor, answer this from your experience as a student.)

Note

1 Mehrabian, Albert. University of California. Quoted in Powerful Presentations, by Jacques Waisvisz, 2014. eBookIt.com.

14

The Smartest Engineer

And Other Difficult Students

One reason training is so enjoyable is the variety of students and personalities one has the opportunity to encounter. Most students are eager to learn, respectful, and professional toward the instructor and other classmates. Technical experts, however, can have very set opinions. They may have spent years becoming more and more proficient in their technology, specific application, or solution. If you are teaching something that challenges their experiences and expertise, you may face some challenging students.

Following are some thoughts on how to handle these situations. But remember, this is a soft skill. These are opinions and opinions are not facts. If these solutions don't work for you, try someone else's opinions! I have grounded these solutions in common sense and adult learning principles. They have proven helpful to many and I hope you will find them helpful as well.

Set the Expectations at the Beginning

Years ago, I was working with a children's group and decided to let them set the rules for engagement. They loved making rules. They were much stricter on themselves than I would have been. They were somewhat disappointed when I narrowed the list of rules from their 20 or 30 items to about 3.

As it turns out, adults are not much different from kids in that regard. If one even hints at turning them loose to make the rules at the beginning of a class, they will likely end up with a fair number of good but trivial rules of classroom behavior.

It is important, however, to agree to certain rules of engagement. Articulate those rules early, before you need them. I tell students very clearly, but in a fun way, that I have only two rules: you must take responsibility for your learning, and you must show respect for other learners.

Take Responsibility for Your Learning

Here are three practical ways you can encourage your students to take responsibility for their learning. First, at the beginning of every class, have each student write down at least one thing they want to do differently after completing your class. This exercise not

Product Training for the Technical Expert: The Art of Developing and Delivering Hands-On Learning,
First Edition. Daniel W. Bixby.
© 2018 John Wiley & Sons Ltd. Published 2018 by John Wiley & Sons Ltd.
Companion website: www.wiley.com/go/Bixby

only encourages responsibility but also emphasizes that everyone can learn at least one thing. Even the know-it-all who would rather be teaching now has at least one thing they will be forced to listen for.

Second, ask them to help you help them. Inform them that taking responsibility of their learning means that they must inform you if something is hindering their learning. The deterrent may be something you are doing, or it may be a hindrance caused by the environment. You may not always be able to fix the problem, but being aware of it helps. More importantly, making them aware of their responsibility to inform you eliminates issues later on.

Third, give them a 2-minute primer on adult education—at least on two points. Tell them that no matter how good of an instructor you are, you cannot learn for them–they must learn for themselves. Inform them that they learn best by doing. "Watching someone else perform a lab or exercise is not enough. You need to do it yourself." Let them know that there are always students who are naturally the first to volunteer when asked. Be grateful for them. Treat them as partners in encouraging others, the more timid students, to participate. Ask the timid students to step out of their comfort zone and prepare them for a class that requires participation.

These three practical tips will lay a groundwork for learning that helps to eliminate personality issues. Addressing these three areas immediately allows you to be free of feeling as if you are singling out any students who might have already demonstrated a particular trait.

Set the expectations of responsibility in both directions. Your students also have expectations of you. Promise them that if they ask a question you cannot answer that you will tell them you don't know, but that you'll write it down and get them an answer. Assure them that you will do your best to keep the class fun and engaging. When students come to training, they are often aware that the instructor's job is to teach them. You need to assure them that you take this responsibility seriously. They are usually less aware of their own responsibility to learn. Taking a few minutes at the beginning of class to clarify those expectations in both directions will create a healthy learning environment.

Be Prepared for Difficult Responses

Setting the expectations is helpful, but you may still get difficult responses, even after the best of introductions. Think through each of the following possible responses before you go to class. It will help you feel prepared for those difficult students.

The Stubborn Mule

"I've already made up my mind."
You may encounter that stubborn individual who has already determined to dislike your product or technology. Perhaps they prefer the competitor's product, or maybe they are late to the new technology and don't want to take the effort to learn anything new. The best way to handle stubborn students is to both acknowledge a difference and minimize the difference. Often, stubborn students just want their opinion vocalized.

They also want you to know that you are not going to change them. If possible, find them alone and, in essence, tell them you didn't expect to change them anyway. Consider a private conversation similar to the following.

> *You know, I noticed you dislike our product. It seems you prefer our competitor's product. Am I right? Well, I'm fine with that. As long as you can show respect to the rest of the class by not interfering with their learning, I don't care. If anyone wants to learn about our product, they have to want to learn about it themselves. My job is to help those who have an open mind learn something new. Are you okay with letting me do that?*

The Pessimist

"This solution won't work."

I was trying to begin a class to a group of particularly unhappy technicians when I got that response. "It won't work." The loud and belligerent comment came from the back of the room. After repeated attempts to begin the class, I finally had the following conversation.

> *You're right. This solution might not work for you. If you have a better solution, we can certainly discuss that later. Right now, my job is to help those who are willing to learn about this product and this technology. If, as you say, it doesn't work, you will not be any less skilled in the technology you are already using. If it turns out that this solution does work, those who learned will be among the smart ones. Who here is willing to take that risk?*

In this particular case, the entire class arrived upset, but no one wanted to be unwilling to learn, like the one who was vocalizing his concern. His overemphasis ended up working against him, as the rest of the class decided they wanted to learn. Incidentally, the solution did work!

The Helper

"Let me do it."

Some students will always be the first to volunteer. That is a good thing, but it can be detrimental to others who want to learn. Your job is to distribute the opportunities fairly. Consider the following options:

- *Who, on this side of the room knows the answer to this question? [Gesturing away from the student in question].*
- *Someone who has not answered a question yet, tell me...*
- *I would like a volunteer that hasn't had an opportunity...*
- *I bet someone from this table can tell me...*

I am sure you can come up with your own, but you get the point. You do not have to single out individuals for a response; you can encourage them by addressing smaller subgroups.

The Talker

"And then there was that time when I..."

Some students love to hear the sound of their own voice. They may be your best student, but are hindering the learning of others or causing the class agenda to derail. "The talker" is the hardest to address because they often do not have an attitude issue; they just love to talk. One approach to take with an excessive talker is to remind them of the need to stay on track. Use nonverbal cues, such as moving away from them (to disengage) or stopping a gesture in midair. They will subconsciously see that you are waiting to go on and it may—if you are lucky—push them to finish quickly. At times, it may be helpful to talk to them during a break and let them come up with a signal between the two of you that indicates a need to get back on track. If "the talker" feels like part of the inner circle, they may be more willing to help you progress. Sometimes you just have to interrupt them.

> *Thank you, but I only have a few minutes to finish my point...*
> *Would it be possible to hear the rest of the story during our break?*

The Extreme Introvert

"..."

Some students are too quiet. They do not want to say anything. You should note that there are levels of timidity. Some only appear to be timid because there are others who are more outgoing in the class, and they don't see the need to interfere with them. Others would be timid even if they were the only one in the class. It can be difficult to interpret a student's level of timidity. Pay careful attention when your trainees introduce themselves at the beginning of class. This is your first and best opportunity to decipher a student's timidity level.

The best thing to do is to draw the timid student out gently. If you have laid out the ground rules, the expectations will be clear: you expect involvement from everyone. Even so, do not push it too hard. Start by involving the more outgoing students and find simple questions to engage the more timid ones. Thank your attendees for answering questions. Thank all of them, since thanking only the timid student will draw unwanted attention to them.

If you have a particularly timid student, one thing that can be effective is to have the students read each bullet point as you progress through the curriculum. Do not start with the timid one, but move down the line and on their turn they can read the next item just as everyone else. These small wins will eventually turn into questions and engagement will become a reality. Never expect too much. Be content with having stretched the timid student just a little.

The Sleeper

"Zzzz"

Almost every class has a daydreamer or sleeper. Get their attention quickly, before they drift into a land too distant for communication! Use nonverbal body language to get their attention. Move into the daydreamer's space without drawing attention to them. Usually, that is all that is necessary. If a student is daydreaming, moving into their personal space is usually sufficient to get their attention. If a student is truly tired, it might be a good idea to take an early break. State something like the following:

> *It looks like we could all use a 5-minute break.*
> *One of the requirements for learning the next section is 5 minutes of fresh air. Let's do that now.*

The Expert

"I already know this."

Probably the most dreaded student in technical training classes is the self-perceived expert. There are two distinct attitudes in an expert. Some experts simply want the instructor to acknowledge them as such. You should. There is no harm in doing so and it will usually help you in the end. Get their assistance teaching other students or find other ways to use their expertise.

Some experts can also be stubborn. This is especially true if you appear to be trampling on a technology in which they have spent years becoming an expert. Jack, a training manager I had the privilege of working with for a few years, was a master at dealing with difficult experts, whom he referred to as "The smartest engineers." He enjoyed the challenge of getting them to change their minds. I remember watching from the back of the room as a particularly difficult engineer challenged him about his product. The conversation went something like this.

> *Thank you for that feedback. Yours is one opinion and it may be correct. Would you be willing to come up after class and help me understand your point a little better? Like you, I always want to learn more. Maybe together we can determine if we really disagree, or if we are just using a different approach.*

The rest took place after the class. I don't know how the conversation progressed, but I do remember that the engineer left having changed his mind. By acknowledging his expertise and being willing to learn from him, Jack was able to individually extend the training session and change his attitude toward the proposed solution.

Conclusion

"It's all about engagement."

As stated earlier, the best way to deal with most of these problem scenarios is to keep your trainees actively engaged in the learning process. If lecture is the only way you are delivering the content, you are welcoming either a debate or a nap. The concern with lecture is that it does not tell the students how they should respond. Students will respond to lecture, but they will respond in whatever way they desire. The individual student is deciding how to respond for themselves, which causes some to sleep or daydream, others to talk, some to argue and be stubborn, some to volunteer to help, a few to check email messages, and so on. By creating engaging activities, the instructor is telling the students how to respond and this type of effective guidance will significantly reduce troubling circumstances.

What if it doesn't work?

Every rule seems to have an exception. Some people are just jerks. Even after setting the expectations, keeping the students involved and actively engaged, and addressing them with respect, some will insist on disrupting the class. In those rare cases, remember these three principles.

1) **Safety first.** If the person is volatile or has been physical at all, you should dismiss the student immediately. Contact his or her supervisor or other authority and seek to get back to learning as soon as possible.
2) **Do not be drawn into their trap.** If they are argumentative, they want you to react. As soon as you do, you become part of the problem. Don't fall for it. Stay calm. Give the class a break and try to talk to them in private.
3) **Do not sacrifice the learning of the rest of the class because of one student.** If it gets bad enough, you may need to talk to their supervisor and you may need to ask the student not to return to class.

In 20 years of training, I have never had to dismiss a student from a class because of behavior. It's possible that I've just gotten lucky. I believe it has everything to do with the methodology of learning by doing.

Making It Practical

Teaching to your peers in the industry can be challenging. If you are in that position, it is because someone believes in your ability to control your classroom and teach effectively.

1) At the beginning of your next class, how will you encourage your students to take responsibility of their learning?

2) How can engaging students reduce conflicts in the classroom?

Before you read Chapter 15, "Virtual Facilitation: Tips for Effective Webinars," answer these two questions.

1) What are some of the biggest challenges in engaging students in a virtual environment?

2) What are some things that should not change, even if the delivery method changes from a classroom environment to a virtual one?

15

Virtual Facilitation

Tips for Effective Webinars

As a modern product expert, you must be an expert in much more than just your company's product. Technology has created multiple ways to communicate with your customers. You may need to deliver a webinar or video training or presentation. If so, you should understand what does not change and what does. Since there are many helps available for those wanting to deliver ad hoc or one-time presentations via a webinar, the focus here is on actual training classes delivered in this format.

What Doesn't Change

Much of what you have read to this point does not change just because the medium you are delivering it in is changing. There are some nuances in each of them that may change, but one should not compromise the guiding principles. It is a good idea to review what those are.

The Philosophical Approach

The way adults learn is the way adults learn, regardless of technology or delivery method. If anything, this will become even more obvious in a virtual setting.

- Proficiency still requires students to do something in order to become proficient. It may take more creativity to make that happen, but it is still part of becoming proficient in something.
- Students will have to take even more ownership and responsibility for their learning.
- Product solution experts will have to be even more adept at thinking consciously about how they do what they do.

It is frightfully easy to fall back into old habits when one directs their focus at something besides learning. Never allow the focus of a webinar be to deliver a webinar. Make sure that your real goal is to deliver effective learning that just happens to be delivered live via a computer and the Internet. Do not let the delivery mechanism force you to compromise the adult learning principles you are learning to apply to instructor-led training.

Product Training for the Technical Expert: The Art of Developing and Delivering Hands-On Learning, First Edition. Daniel W. Bixby.
© 2018 John Wiley & Sons Ltd. Published 2018 by John Wiley & Sons Ltd.
Companion website: www.wiley.com/go/Bixby

The Structure

Use the 4×8 Proficiency Design Model to design webinars. While the process is the same, it is usually wise to separate out the design process of a webinar and a face-to-face, instructor-led class. You cannot teach a class designed for the classroom over a webinar or video and expect the same results.

Remember, in the best-case scenario, a curriculum designer should not choose the delivery format until step 6 of the 4×8 process. In rare cases where the delivery format is pre-determined, you will need to go through the other steps with that in mind.

- The business goal. What is the goal for this webinar? Keep it simple, but you must have one.
- The intended audience. Who are they? Even though you cannot see your audience, you must define them.
- The objectives. A good objective must still have an observable and measurable verb. You may need to change the verb, since the action must be observable virtually. Unless you are using cameras at the user's end, you will not be able to observe them installing hardware, so you will need to rewrite the objective to include something you can observe. In a webinar, it is best if you can reduce the training to one main objective.
- The outline. The process for creating the outline is the same.
- The activities. Students will need to perform the activities virtually. You may decide to have these done outside of the webinar and turned in separately.
- The delivery method. Even though you have already determined the delivery method, you may choose to enhance your training via video, a pre-requisite reading, or some other learning delivery.
- The assessment. You can complete the assessment online through polling tools built into the system or through verbal acknowledgments if the class is not part of a certification program. Students can also complete the assessment separately from the webinar.
- The content. Create the content specifically for the type of delivery that the instructor will use. Use the capabilities of the tool. Most tools have the ability to create polling questions, add attachments, and so on. Learn the tool well and use it fully.

The Definition

A presentation delivered via webinar is still a presentation. Training is still about changing what people do. This may restrict you from delivering via webinar what you thought you could. That is fine because that means when you do, your webinars will be more effective.

Facilitating Virtually

You do not want only to deliver content. You know that you need to help adults learn by doing, but the environment challenges you. Take heart; you are not alone. Proficiency training in a virtual setting does have its challenges. Often, one must deliver the virtual training as part of a blended approach, to allow students to perform tasks in a hands-on training environment.

There are times when you can use a virtual environment to help students become proficient in a product or solution. Your goal is to deliver effective learning. Here are a few tips.

Regarding the Presentation

1) **Use even fewer words on your slides.** You want to keep your students' attention. If they can get everything by reading the slides, they will do so and then start checking emails. Do not give them that option or frustrate them, wishing you would have just sent them the slides to read.
2) **Use even larger font sizes.** Remember that some students will be using a mobile device. Make sure one can read the text from a small device.
3) **Use lower resolution pictures and graphics.** You do not want students with a low-speed connection to get behind.
4) **Reduce or eliminate animation.** Animation rarely works well in a webinar setting. Audiences connect to your webinar over different networks with different speeds and bandwidth. The simplest fix is to eliminate animation altogether. If you must use it, use it sparingly.
5) **Create a welcome slide.** Start with a welcome slide that includes extra dial-in information, a reminder to mute phones and not to put phones on hold, and any other helpful information. Start the webinar 5–10 minutes early with this information.

Regarding the Tool

1) **Learn to use the tool well.** Practice how to answer questions, mute and unmute phones, give control to another presenter, upload documents, take a poll, and so on. All of these items will come in handy.
2) **Upload the presentation into the tool.** Most webinar tools allow you to upload the presentation prior to showing it. Do not attempt to give a webinar by simply showing your screen. This slows the connection down significantly.
3) **If possible, use a camera.** It is easier to keep your students' attention when they can see you. Ideally, you could use a tool where you can see them as well, but if that is not available at least try to let them see you.

About the Event

1) **Keep the class size small.** Just because the delivery format is not in a classroom does not change the fact that training requires a smaller audience. You are still teaching individuals. Know how many you can handle for the objectives you want to meet and stick to that size.
2) **Use an assistant or moderator.** You may think you can handle it, but you will not regret having help. I have done many webinars in my career, and I usually use a second person or moderator. I need the extra hands to help with technology, questions, and any issues that may arise. An assistant can be extremely helpful even if they are not physically present with you.
3) **Run a pilot class.** Always run a pilot class, just as you would any other class you are offering for the first time. Doing so will help to decrease issues in later classes. Get as many variations as possible to view and interact in the webinar. Use multiple browsers, multiple mobile devices and operating systems—all with the goal of knowing how to make your webinar as effective as possible.
4) **Engage the students.** The format is not an excuse for not engaging the students. Most good tools offer several ways to do that.

- Polling questions. A good webinar software program will integrate a live polling option into the system. Instead of just asking questions verbally, ask the questions in written format as well and let the students answer live.
- Whiteboard options. Some software programs allow students to write answers to questions or click on a graphic question.
- Use a seating arrangement. Showing a "seating" arrangement proves that the instructor is aware of who is attending the class. You may also want to print it and keep it by you. When you ask questions, you can tell them that you are going to go in order, or you can choose students to answer based on their seating assignment. It also helps to make the students feel like they have to stay involved in the class.

5) **Use good nonverbal skills.** Good verbal skills are obviously important in a webinar. What many forget is that good nonverbal skills are still important, even if your students cannot see you. Try standing when you are giving a webinar. It will increase your energy, and it will be easier for students to remain engaged. Remember to smile. They can hear you smiling!

Conclusion

Technology is a wonderful tool and can simplify the way you communicate with students around the world. The way one delivers information, however, does not change the way adults learn. Never assume that telling students something, or showing them a few slides, equates to good training. Engaging your students and making sure you meet each objective is still critical. Find ways to let your students perform tasks that will help them internalize the learning. Your students cannot be proficient in something they have never done, but with a little creativity, you may find ways to make that happen at a distance.

Making It Practical

The way you deliver training does not change the way adults learn. Never allow the delivery method to make your training less effective.

1) What is one thing you are going to do differently next time you teach via webinar?

2) Should you use the 4 × 8 Proficiency Design Model to create an effective webinar? Why or why not?

Before you read Chapter 16, "Technical Presentations: Effectively Design and Deliver Technical Information," answer these two questions.

1) In your own words, what is the difference between a training course and a presentation?

2) What is one reason you might decide to offer a technical presentation on your product?

16

Technical Presentations

Effectively Design and Deliver Technical Information

With so much emphasis on *doing*, it may seem that *knowing* is not important. The reality is, of course, that knowing and receiving information is extremely important. Sometimes that is what your goal is. In fact, some of you may have been disappointed to learn, but as you read this book that much of what you have called training in the past is, in fact, a presentation. Your task has been to convey knowledge, not to help improve another individual's skill.

It may be preferable to transform your presentation into a true training class. I hope that many of you will identify this gap and change your presentations into effective and experiential learning. But presentations are not going to go away, and it is important to get them right as well.

Presentations that are designed correctly can help to build a strong foundation for future training. Most training classes have at least a few presentations sprinkled in them and most product specialists will deliver more technical presentations than they do training classes. When these presentations are related to products, however, the goal is still to change behavior—or at least provide information that can lead to a change in behavior when that information is applied.

Many of the training concepts described throughout this book can be applied to technical presentations, with a few notable exceptions. While a presentation does not generally include hands-on learning, it can be an effective prerequisite to experiential training.

When to Use Presentations

In Chapter 6, I offered Table 16.1 as a way to emphasize that training is different than a presentation. The chart emphasizes the value of training in regard to teaching a skill or changing how an individual performs a task.

Notice, however, that all of the things listed in the *presentation* column are good things! Many are important to do. Take a closer look at the positive reasons for delivering an effective technical presentation.

Product Training for the Technical Expert: The Art of Developing and Delivering Hands-On Learning, First Edition. Daniel W. Bixby.
© 2018 John Wiley & Sons Ltd. Published 2018 by John Wiley & Sons Ltd.
Companion website: www.wiley.com/go/Bixby

Table 16.1 Presentation versus training chart.

Presentation	Training
• Successful when it changes or enhances knowledge—about knowing	• Successful when it changes or enhances a skill—about doing
• Motivates and encourages change	• Demonstrates change
• Audience size is irrelevant	• Learning is individualized
• Validation can be immediate	• Validation may be lengthy
• Focus is on the delivery; how it is *presented*, and builds on the presenter's experience	• Focus is on the receiving; how the learning is internalized and builds on the student's experience
• Often *is* the learning event	• Generally is one step in a process

When the Objective Is to Deliver Information

A presentation is successful when it changes or enhances knowledge—and changing and enhancing knowledge is important to do. Information is necessary in order to change knowledge. There are many ways to deliver information, but a presentation has advantages that a mere document or even a video does not. Just like face-to-face training, it has the powerful element of human interaction.

When information is coupled with a presenter's experience—their stories, their expressions, and their ability to tailor it to a specific audience—it becomes more powerful. Books are helpful, videos are powerful, but a good speaker can capture attention and deliver information more powerfully than either of those methods.

There are times when that is all that is required. There are times when you need to deliver important product information, and changing a skill is not required. There are also times when familiarity or urgency requires this information to be delivered in a setting that requires a presentation. In these cases, the goal is to give a presentation that will *eventually* change behavior.

When Time Is Limited

Sometimes there are real and unchangeable time constraints that will require information to be presented via a lecture or presentation. Information, of course, can be delivered most effectively in a proficiency training setting. However, when time limits remain the same, many new instructors are surprised at how much material they have to eliminate in order to deliver hands-on learning. Most trainers who do much presenting are used to delivering significantly more information in less time. In most of these cases, the right thing to do is to convert that presentation into a true training class and increase the time required to do so. Experience has taught me that it generally requires at least twice as much time to teach something well as it does to present the material. However, that is not always a practical option.

While it is important to limit the amount of material delivered in a presentation (many present too much at once), the material you use in a presentation can vary significantly from your training content. A good presenter can more quickly change a listener's knowledge than even a great trainer can change a student's skill.

When time is limited, it can be helpful to take a hybrid approach. Use presentations, either in person or virtual, to deliver prerequisite information and follow up with hands-on exercises or a list of projects that the student can do. This approach should be used sparingly, because the learning from the presentations will be minimal, unless the student is already an expert.

When the Audience Is Large

Presentations are often delivered to a larger and broader audience than hands-on training classes. Audiences with varying degrees of expertise require the presenter to address a given topic in broader terms. Often, the learning must encompass enough background information to help a beginner and enough advanced material to benefit an expert. These presentations often *are* the learning event—they are not part of a systematic learning curriculum—so they must present an entire package in a concise format.

Most public speakers deliver a significant amount of material and challenge the listener to choose how they will apply that information. No speaker expects any listener to remember 100% of what they hear. Attendees with advanced skills are going to capture more information and apply that information differently than beginners will. This is why presenters who use lecture as their only method of teaching get frustrated at the lack of practical results. They assume that listeners will apply the learning in the same way they themselves, as experts, would apply it. When their students apply it differently, they get confused. The reality is that their students may be applying it differently because they learned different things. This is the beauty—and the danger—of a one-way delivery method.

There are two ways to overcome the problem of multiple applications. The first is to embrace it and encourage it. Depending on what you are teaching, this is a natural and positive response to many presentations. It is the ambiguity of the objectives of a lecture that make it a great tool for stimulating creativity. While training focuses on how to do something specifically, presentations focus on how to know something generally. Both are important, as long as one isn't used exclusively to do the other. In other words, you can't use training exclusively to teach general knowledge, and you can't use presentations as the exclusive way to teach a skill.

The second way to overcome the problem of multiple applications is to make the presentation so simple and straightforward that there is only one application. This is difficult for most presenters to do. It means that your presentation must be short and have only one or two simple applications. These simple learning sessions are often referred to as "bite-size learning" or "micro-learning." Providing small bits of information at a time is effective because the brain only needs to remember one or two small details in order to put it into practice. The problem with this approach is that it doesn't take into consideration the various levels of the audience. The application you choose may be great for one listener, but not for another. When that happens you lose them as an intentional listener.

This method works well in short, recorded presentations, however. The reason is that the listener is choosing this application to learn about; it isn't being forced on them by the instructor. I encourage micro-learning for peer-to-peer training and learning that is meant to serve as continuous improvement for experts.

Table 16.2 Four learner-readiness principles.

Principle	Question they must be able to answer
Learners must recognize the need for learning	Why do I need to learn it?
Learners must take responsibility for their learning	Will I put forth the effort to learn it?
Learners must be able to relate the learning to their experience	Can I relate it to what I already know?
Learners must be ready to apply it	When will I need to apply it?

Some large audiences are comprised exclusively of experts at similarly advanced levels. A large conference of physicians or electrical engineers, for example, may have less of a problem with varying skill levels, but is too large to teach in an experiential manner. Lectures can be meaningful to these audiences precisely because of the similarity of background and education. Presenters can often get very technical because the receivers of the information are at similar expertise levels as the deliverer of the information. The assumptions that the deliverer is making are probably more accurate, which makes the information more attainable and more likely to be absorbed.

In all of these cases, however, the intention is still to convey knowledge, not to perfect a skill. Doctors attending a conference must still take the knowledge they learn and apply it in their practice. The same is true of engineers or product specialists.

To Motivate and Encourage Change

One powerful reason to use a presentation is to motivate or encourage change. In Chapter 4, I covered the four principles of learner readiness. When ability is equal, these principles will influence how much and how quickly a student will learn. Look at the four principles in Table 16.2 from the perspective of a technical presentation.

Presentations are a great way to prepare a student to learn. An experienced presenter can use a technical presentation to convince students that they need to learn more or perfect a particular skill. A captivating speaker can use a technical presentation to motivate a student to give their best effort toward learning. A good storyteller can use a technical presentation to help students relate what they have learned or will learn to what they already know. A good communicator can prepare for training by helping students consider when and how they might apply the learning.

Since these four elements are critical to effective learning, presentations can play an important role in the success of product training. It is important then to design your presentations effectively.

How to Design Effective Technical Presentations

The 4×8 Proficiency Design Model introduced in Chapter 9 can be modified to design an effective presentation. When developing a presentation, only five of the steps are required, though each of those has slight differences (Figure 16.1).

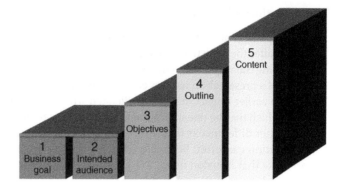

Figure 16.1 Five-step technical presentation design process.

1) **Business goal**

Defining a business goal is always important. This is usually a very simple process when one is creating a technical presentation. The goal doesn't need to be long. Unlike training, the goal does not necessarily need to be measurable. For example, you may want to do a technical presentation just to have a presence or demonstrate thought leadership. It may be as simple as getting in front of a customer for the first time.

Even so, it is important to know why doing the presentation is helpful to your business or product line. Sometimes just thinking about why you are doing a presentation can open up other opportunities or ideas for improvements. Often, they lead to developing effective training as well.

2) **Intended audience**

Like the business goal, stating the intended audience is important, but may be much broader. Knowing if you are addressing experienced engineers or interns should change your objectives. More importantly, it will affect how you achieve those objectives. All presenters make assumptions about their audience. You can't give all the background necessary to learn about every topic. Nor should you. Doing so would bore most audiences. However, it is also unhelpful to assume too much knowledge. Instead of boredom, audiences are left wondering what you're talking about or feeling left behind.

No two audiences are exactly the same. No two presentations should be exactly the same. Do your best to get as clear of an understanding of the typical audience before you create the presentation. Later, when it comes time to deliver it to a new audience, you should verify that the audience, or at least any assumption you are making about the audience, hasn't changed. If it has, but only slightly, you can likely make the appropriate changes in the delivery of your technical lecture. If, however, the audience has significantly changed, it may require rewriting the objectives.

I recently asked an engineer to create a presentation about a topic he was very familiar with. His target audience was new employees—college graduates with no previous experience. His first response was that it would be simple; he had delivered the presentation many times and was confident it would be easy to do. However, what he failed to account for was that his previous deliveries were to other engineering experts. Recreating the presentation for subject matter beginners proved to be much more difficult than he had anticipated.

Never assume you can take the same content and recycle it for any audience. Take the time to get it right. Otherwise, you are wasting your time.

Determine the Delivery Method (Optional)

A quick perusal of Chapter 9 will remind you that I advocate determining the delivery method after the objectives and educational activities have been determined—when one is designing training. Designing a technical presentation is different. Because the delivery format of a presentation has less effect on the objectives, it is fine to determine its delivery at an earlier stage. Often this is dictated when the request for the presentation was made (which may be one indication that it is a presentation, not a true training class).

Another difference is that this step, at least at this point, is completely optional. A presentation can often be delivered in more than one format and you may, or may not, make that decision now.

3) **Set the objectives**

Informational Objectives

Writing clear objective is very important, even for a presentation. In fact, the best presentations have clear, actionable objectives, similar to training objectives. They should still be both observable and measurable. The difference is that the instructor may not have the opportunity to observe and measure that those objectives were actually met. For example, you may determine that you want the audience to be able to define three differences between product A and product B. Stating that clear objective will help you design a lecture that emphasizes and simplifies those three differences. You should make it so clear that any normal person will leave that lecture and be able to state the differences, even if they never had to do so in class.

Because presentations tend to be more about distributing knowledge and less about enhancing a skill, it is easy to get careless with the objectives. Avoid vague or unmeasurable objectives. Instead of writing that you want the student to "know more about your product," write that you want to the students to "list 5 things about the product." Clarity is strength. The clearer your objective, the more powerful your presentation.

Motivational Objectives

Presentations, even technical ones, offer an opportunity that is less obvious in technical training. Namely, they are great for increasing motivation or inspiring change. Motivational objectives are both easier to write and harder to measure. They are more emotional and their influence is often short-lived. You may want to challenge listeners to make quality a top priority or to incite more sales of a particular brand or product. Both are good uses of a powerful presentation.

Even with motivational objectives, however, clarity will make your presentation more powerful. If you state clearly how your listeners can make quality a top priority or specifically describe how and why one product should be emphasized over another, the presentation will be more successful. The fact that you aren't teaching a skill is never an excuse for a vague objective.

Writing objectives for a presentation (or training class, for that matter) can sometimes influence one to change the intended outcome from a presentation to a training class. You may set out to create a presentation and realize that you really need to teach a skill. You may set out saying that you want to create a training class and then realize that a good technical presentation will do the trick.

For example, you may want to challenge listeners to choose the right product for the right job and assume you can do so in a presentation. After you begin to write the objectives, however, you realize that the majority of students don't have enough experience with the products in question to make that decision. It would be a waste of time to motivate them to do something they cannot do. When that happens, change it to a training class. The first two design levels are the same—you are still on track!

4) **Create the outline**

After the objectives are clearly defined you can write the presentation outline. As you write the outline, refer back to both your objectives and your intended audience. For each new topic, write down what you expect the listener to already know. Make sure that what you write matches the audience. This will help to avoid leaving information gaps in your outline or adding unnecessary material.

There are many help available for making a good outline. Most outlines have at least three parts. The first part, an introduction, prepares the listener to learn. This is important because of the learner-readiness principles discussed in Chapter 4. You need to tell your listeners what they are going to learn and why it is important for them.

The second and longest section, often referred to as the body, is where you give out the information or inspiration that you want. The body of a technical presentation is often a systematic layer of information that must be given in a specific order. Most technical experts find this easy to put together, since they like to think in systematic and organized sequences anyway. The difficulty is not adding the necessary information. The problem is subtracting the unnecessary information.

Not all information is equal. Your first outline is a draft that needs revision. You aren't done just because you have numbers and letters or bullet points. Always go back and revise it, comparing each line item to your intended audience and objective. Delete anything that does not help to meet one of your objectives. The next step is to differentiate between content that is absolutely required and material that is nice to know. In my outline, I put a plus (+) sign next to anything that is required and a minus (–) sign next to anything else. A minus sign means that I could still meet the objective even if I don't cover that material. This is especially helpful when time constraints require flexibility. You already know what can be cut or shortened and what cannot be eliminated.

The last part of a good outline includes the closing section. Always leave time to wrap up your presentation with a reminder of your main objectives and a call to action. An effective presentation does not allow the listener to doubt the speaker's intention nor does it leave the hearer wondering what they should do with the information. Remind them. Keep it short and simple enough to leave the room and repeat it. Take all the information and squeeze it into a simple and succinct call to action.

Unlike training, presentations do not always have specific activities that help the learner to construct the knowledge for themselves. However, your presentation should be engaging nonetheless. As you create your outline, note some ways you can engage the audience in order to drive home your point. You may come up with more than one way, so you can vary it depending on the audience size or background.

5) **Create the content**

Once your outline is ready you can start putting your content together. Create content that specifically addresses your outline. Not more, not less. If you are using a visual aid, like PowerPoint, be sure to follow the advice in Chapter 10. Your slides should only be a tool to help communicate; they should not contain all of the information you are delivering.

As with training, the temptation will be to pull existing content and make it fit. Instead, start with the outline you created and find or create content to match it. Use graphics and fill in the information with verbal communication. Only use text to emphasize key things you want them to remember.

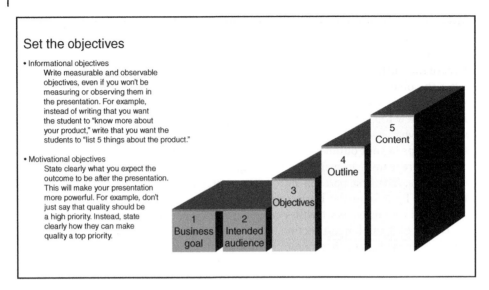

Figure 16.2 Slide with too much text.

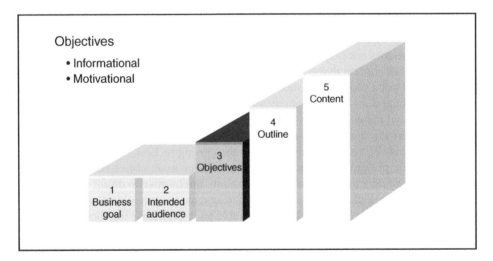

Figure 16.3 Simplified slide.

In Figure 16.2 the slide has too much text. The audience will be encouraged to read the entire text and so will the speaker. The pictures may or may not be helpful, but the listeners will have little time to look at them if they are reading the text.

The brain functions best when it does not have to multitask. While some people are better at multitasking than others, we should avoid forcing our audience to do two or even three things at once. If you ask them to look at pictures, read the text and listen to you, their comprehension level is likely to decrease. Instead of writing out long paragraphs of text, try something different.

In the second example in Figure 16.3, you can still state the facts, but the brain can more quickly process both the text and the picture, which is available only to provide a visual context for the listener.

If you are concerned about distributing information to the students after the class or lecture, put the information in the notes section, not in the slide you are using as a visual aid.

Delivering Your Presentation

Most of the principles outlined in Chapters 11 and 13 apply to giving presentations and facilitating training equally. The biggest difference will be in how you engage your audience.

While there are tips and tricks that will help you engage more, the truth is that every audience is unique. The size of the audience, their readiness to learn from you, their previous experience with the topic you are presenting, and any number of biases will affect how you can engage them comfortably. But you must engage them and the more you know about them, the more natural your interactions with them will seem.

Always talk with the audience, not at them. When the lights are bright and the audience is large, it is easy to turn your presentation into a theater performance. As soon as they feel like a spectator and not a participant, you have lost them. Instead, relax and converse with them. All of them. Turn toward them and speak directly to individuals. They will communicate back, if you let them.

Ask Questions

One way to engage any audience is to ask questions. In smaller audiences you can use questions in the same way you would in a training class. Larger audiences require that you adjust your questions to make sure that you get the response you want.

Unless you know the audience extremely well, it is best to tell the audience how you want them to respond to your question. "Raise your hand if you work for a technology company" is better than "How many of you work for a technology company?" Telling the audience how to respond also creates movement and engagement and doesn't allow anyone to use ambiguity as an excuse for not getting involved.

Of course, asking for a show of hands only gets the audience involved if they raise their hands! So be careful with your questions. Don't ask a question that risks falling flat. For example, if you are presenting about an opportunity to sell a particular product and the room is full of people who sell for a competitor, it may not be wise to ask, "How many here want to hear about this product?" Instead, ask, "Raise your hand if you like to make money." The chances of getting hands in the air with the second question are significantly higher!

When you ask a question of your audience, even a rhetorical question, give them time to answer the question. This may seem odd, since few people will answer the question out loud and you likely don't expect that. But if you are having a conversation with them, you want to give them the time to answer the question in their own minds. Moving on too quickly will send the signal that you don't want them to answer the question and they won't, even to themselves.

Speakers who ask questions of their audiences demonstrate a level of comfort with their listeners and with their topic. You may not be any more comfortable, but you will sound like you are. As your audience engages with you, it will energize you as well and you will find yourself feeling more comfortable addressing them.

Practice, Practice, Practice

You cannot engage an audience if you don't know your topic thoroughly. Whenever possible, give a test run with someone you can trust to tell you the truth. Find out as much as you can about the audience and the venue. Prioritize your presentation so that, if it was reduced to three minutes you would be able to state your main objectives clearly.

Practice the nonverbal communication skills addressed in Chapter 13. Well before the presentation, choose 1 of the 10 verbal communication skills addressed in Chapter 11 that you want to improve on and get some help on that communication area.

Look for other opportunities to improve your public speaking. No matter how much experience you have, you can improve. Chances are your community has resources that can provide opportunities for development. Look for them and take advantage of them.

Relax and Have Fun!

All of this requires you to challenge yourself and challenge your audience. But don't forget to relax and have fun. When you get up to speak, you are no longer practicing. Now is not the time to count your own verbal miscues or panic that you didn't practice your gestures enough. Now is the time to rely on the training you have given yourself. Just as it is with teaching, your job is not really to deliver a great presentation. Your job is to be effective. If your listeners go away with a different idea, a new vision, or an improved understanding of a particular topic, you have done your job—even if you said too many "ums" or forgot to use enough gestures.

Conclusion

Effective presentations—even technical ones—come from the heart. The most effective presenters believe what they are saying and want others to as well. A technical presentation is like a screwdriver in the hands of a subject matter expert and a power driver in the hands of passion. Don't give a presentation unless you believe in what you are saying. Deliver your message, not because you love yourself or the prestige of public acceptance. Give a presentation because you love to see others learn and improve their lives. When that passion is communicated, it will be effective.

Making It Practical

Many times, subject matter experts do not have the luxury of delivering hands-on learning. When a presentation is all that is expected or required, delivering it effectively is important.

1) You will often get requests for "training" that are really requests for a technical presentation. Can you define some of the differences?

2) List the five steps required to create an effective technical presentation.

Before you read Chapter 17, "Culture and Proficiency: Training for Proficiency in a Global Environment," answer these two questions.

1) Does culture change the way people learn a skill? Does it change the way a skill is taught?

2) Can an instructor from one culture effectively train students from another culture?

17

Culture and Proficiency

Training for Proficiency in a Global Environment

I love to see the world. I am fortunate to have lived in multiple cultures and to speak more than one language. I have lived for at least a decade of my life on three different continents and have traveled and taught around the globe. This experience, even though some of it was as a young child, has been very helpful to make me effective as a trainer around the world, even though it does not make me an expert in any one of those cultures. In fact, sometimes I feel more like a man without a culture than one experienced in many of them. What the opportunity to live around the world has done, however, helps me to approach culture more openly and neutrally. I still get culture shock like anyone else. I am just a little bit more prepared for the jolt. You can be prepared as well.

I love learning about culture because I love observing how best to teach people. As I have traveled, I have seen that learning happens in the same way, but teaching can vary from culture to culture. Learning is part of being human. It does not change. Teaching, on the other hand, is communication. It does change.

This book does not explore the relationship between culture and learning. It is about helping you, a product expert, be effective when you train others on your products. If you are lucky, you may have the opportunity to do that in a culture that is foreign to you. If so, an awareness of what to look for and what to expect can be helpful. There are things that you should not change, and there are things you should be willing to change.

When you go to a different culture, you are a learner, and the best way to learn is by doing. Be humble. You will make mistakes. Listen to your students and learn how to teach them. That is what proficiency instructors do even in their own culture. It is the best way to help others learn. Embrace the opportunity and the challenges (because there will be those as well) and continue to be effective as an instructor.

Even if you never have the opportunity to travel to a different culture, you will likely teach students with different cultural backgrounds. Figuratively speaking, the world is much smaller than it has ever been. The following tips and suggestions are helpful even to trainers who don't travel.

Product Training for the Technical Expert: The Art of Developing and Delivering Hands-On Learning,
First Edition. Daniel W. Bixby.
© 2018 John Wiley & Sons Ltd. Published 2018 by John Wiley & Sons Ltd.
Companion website: www.wiley.com/go/Bixby

What Doesn't Change

People are people. There are certain things about humanity that do not change, no matter one's skin color, what part of the world one lives in, or what religion or nonreligion one adheres to. The way adults learn is one of them.

The Philosophy of Hands-on Learning

No matter where you are on the globe, adults learn to be proficient by doing. That approach to teaching shouldn't change just because your training location changes. In fact, you will often find it better to reduce lecture *even more* and increase hands-on learning when you travel, especially if you don't speak their language. Language is only an issue when you are speaking. It is not an issue when they are doing.

Do not confuse expectations with culture, and do not confuse education systems or even education norms with effectiveness. Adults in different cultures may be used to learning in different ways. For example, some cultures are extremely competitive and their teaching techniques rely heavily on memorization. Most business cultures are familiar with presentations as the primary method of transferring knowledge—even though presentations are an ineffective way to teach a skill. What adults are used to in a formal education setting and what works may be two different things. That adults are used to technical training being a boring PowerPoint presentation doesn't render it effective, no matter what culture you live in.

The principles of learner-readiness are also cross-cultural, though they may show themselves differently. Students must still recognize their need for learning, they must take responsibility for their learning, they must be able to relate their learning to similar experiences, and they must be ready to apply it.

Taking responsibility for learning can change its focus in different cultures. While individual personalities are still relevant, regional culture may enhance certain personality traits. In some cultures, a successful student will sense a strong responsibility to learn what the instructor is teaching, while other cultures may emphasize a stronger responsibility to make sure the instructor is correct. Students from certain cultures may display a strong sense of responsibility to protect the instructor's honor and will rarely challenge him or her. Students from other cultures may openly disagree, or even appear to argue with the instructor, out of a sense of responsibility to get all the facts, even if the facts are unfavorable. When you understand that your students are exercising their responsibility to learn, albeit in different ways, it makes the training event more enjoyable and more effective.

One of the benefits to learning while doing, as opposed to learning while listening, is that active learning tends to neutralize cultural communication differences. When a student is completing an exercise or lab, the communication between them and the instructor is minimized, reducing opportunities for miscommunication.

The Strategy of Hands-on Learning

The strategy of product proficiency training is also consistent across cultures. Selling more of your product is your ultimate goal, no matter what hemisphere you live in. The expectations of what great training versus adequate training is may change, but the effort to make it great is still required.

Assessing the effectiveness of the training is also still important. Taking a more objective approach to collecting feedback will increase the value of the information you get regardless of the culture in which you are assessing. Objective feedback is even more valuable in cultures where the default is to give high marks regardless of the quality of training.

The Structural Design of Hands-on Learning

For the most part, the structure of your training should remain the same cross-culturally. You should continue to make content more important by getting away from the content-first approach and using the 4×8 Proficiency Design Model. Since the model builds on adult learning principles, it should not change. You may need to alter certain parts of it, however, such as the exercises you use to teach certain objectives. Feel free to regionalize the exercises, as long as they continue to demonstrate proficiency of the objective.

If the objectives themselves need to change, you may need to create a regionally specific training class or regional product certification program. Use an 80/20 rule. Allow for 20% variation of the class and regional customization before creating an entirely new class.

What Does Change

Some change is inevitable. Training is a communication art, and culture affects communication, even if the principles of learning do not.

The Delivery of Hands-on Learning

Because the visual aids and student guides are an essential part of the communication process, they may need to change as you deliver the training in different cultures. Here are few things to consider in regard to your PowerPoint presentation and student guide:

Include regionally understood graphics and pictures. Take the time to demonstrate a desire to regionalize the learning by using graphics that are meaningful to your audience. No audience should feel like they are a secondary audience from a cultural or geographical perspective. You carefully targeted a particular audience based on your business needs. It is likely that there will already be some from outside of that scope in your class. Don't unnecessarily create even more secondary audiences based on culture alone.

Use regional case studies and examples. The purpose of regionally specific examples and art has an adult learning foundation in the four learner-readiness principles. Bring the illustrations and examples closer to the student so as to help them answer the questions "Why do I need to learn it?" and "Can I relate it to what I already know?" Regionalizing your examples demonstrates a desire on your part to deliver learning that matters to the local student, not just a desire to deliver content.

Use even more pictures. There is a tendency to use more text when facilitating an international audience. There is a place for more text. It is not on your PowerPoint slides. Include the text in your notes, but remember that your students may understand pictures much more easily than they understand text.

Create a more detailed student guide. Do not expect students to be able to take as many notes if you are not speaking their first language. Put more information in the student guide. Most students will read a second language better than they can speak or audibly understand it.

Whenever possible, solicit local help to review content. It is always a good idea to ask someone locally to review your content. Students from large, international corporations may be more comfortable with global content or examples, case studies, and graphics from faraway places. The best way to know how to be effective with your particular audience is to ask a representative prior to the class.

The Facilitation of Hands-on Learning

Facilitating, or making learning easier, must happen in any culture. The way you make the learning easier may change, however. You must be flexible in how you interact with your students while making sure that you meet the objectives for the training. Here are a few tips for training students who do not speak English as their first language:

Slow down. Yes, you should speak slower, but you should also slow down in general. Take more breaks. Plan for more time to get through the material. This is especially important, of course, if you are working through an interpreter. Even if you are not, you should plan for more interaction, more explanation, and more time to absorb the content.

Avoid idioms or trendy phrases. I still have to work on this. Recently, I was teaching overseas when I asked students to write down one *takeaway* from the class. After several incredulous stares, one brave student finally asked for an example. I gave an example of what I meant by "takeaway"—something the students wanted to learn or do better as a result of coming to the class. After I gave the example, she smiled and said, "Takeaway means *to take away*. We thought you were asking us what we wanted to remove from the objectives." Lesson learned.

Engage, engage, engage. I hope that by now you understand the importance of engaging your students. If your students are learning in a second language, however, engagement is even more important, because it also improves comprehension, not just learning.

- Make the directions for your exercises clear and easy to understand.
- If needed, teach the exercises to one of the students beforehand and let them help explain them to other students.
- Have a backup exercise in the event that one doesn't work well with your audience.

If at first you don't succeed, try and try again. Just because you told them doesn't mean you have trained them. This is true in any language or culture, but especially important to remember when you are training students who are learning in a second language. Do not move on until they have learned what you are teaching them. This will require asking many questions and letting them practice even more. Don't look at it as their only opportunity to hear what you have to say about your product. In reality, this may be your only opportunity to make certain they know how to do something with your product.

> **A note to trainers teaching in a second language…**
>
> First, congratulations on your ability to teach in more than one language. The tips I noted earlier and throughout this book will increase your effectiveness even if you are teaching in a second language.
>
> As students learn while doing, you will need to speak less. Even when you do have to explain and present theories, the fact that you are facilitating a conversation and not giving a speech will decrease anxiety and make the experience more enjoyable for you and more effective for your students.

Other Tips for the Traveling Trainer

Training in different cultures requires flexibility and humility. You will have to adjust your communication techniques for cultural expectations and approaches. Do so without compromising your philosophy of learning while doing. Learning is constant, even if teaching methods are not. Know that there will be challenges and you will certainly face unique circumstances.

Here are a few tips to consider before training cross-culturally:

Do your homework. Take the time to read up on the country and area that you are visiting. Earlier, I recommend Morrison and Conaway's book, *Kiss, Bow, or Shake Hands* as an excellent resource. Knowing your audience is important—as is knowing their cultural background. You cannot expect to know everything, but you will gain their respect and their attention if you have spent a little time learning about their culture before you go.

Admit what you don't know and embrace the differences. No matter how much homework you do, you will still be a foreigner. I have avoided many awkward moments by being honest about my lack of cultural knowledge. Ask your students to help you. They will enjoy the opportunity to train you! When you refer to the "homework" I mentioned earlier, do so with questions, not dogmatically or stereotypically. Culture is dynamic and your students want primarily to know that you value them as human beings and that you are interested in their culture because you are interested in them. Enjoy the cultural opportunities you get. Ask a lot of questions. Try new foods. Experience new things.

Arrive early. I know I said that it is important to be the first person to the training event earlier in the book. Here, I am not talking about an hour; I am talking about a day. It is important to be as rested as possible, even if you can't expect to completely adjust to the time difference. Arrive in time to deal with potential problems or necessary changes. Schedule international training loosely enough to allow for travel changes and other adjustments.

Conclusion

One of the best ways to learn about a culture is to teach people in that culture. Teach—not give a presentation. When you engage with your students, when you allow them to ask you questions and you see how they interact with you, with their peers and with your product, you can learn a lot from them. Few people have this wonderful opportunity. If you are one of them, embrace it. Learn and teach at the same time. You will grow professionally and personally, and it will make you a better trainer no matter where you teach.

Making It Practical

Traveling can be fun. Knowing that you are changing how people use your products and solutions is even more rewarding.

1) How does learning by doing minimize cultural differences?

2) Why is it important to use regionally specific and understood case studies and examples?

3) How does teaching in a different culture increase the importance of listening to your students?

Review of Part Four: The Facilitation of Hands-On Learning

1) What is the main difference between lecturing and facilitating? How does that affect how your students learn?

2) What are you going to do differently next time you teach, as a result of reading Chapters 11–17?

Part V

The Operation of Hands-On Learning

18

Certifying Proficiency

The Fundamentals of a Product Proficiency Certification Program

You have put a lot of effort into creating an effective training program. Now, you may need an official verification that an individual can use your product. You can do this with a certification program. Your need for a certification program may be customer driven, but often the need is driven by a desire to identify resellers or end users that are trained in how to perform specific functions with a product. Certification may exist to verify a number of skills. One may confirm that a person is able to specify or design, while another may declare one competent to install or merely use a product.

Because people change occupations much more frequently today than they did 40 years ago, certification is a popular way of putting the responsibility to educate users on the manufacturer of the product. Manufacturers offer proficiency certification programs to individuals who have completed certain requirements that demonstrate competency in a product or a proprietary solution or technology.

This chapter does not offer a legal definition of certification nor does it legally define the requirements. However, there are some general concepts that one should understand before creating a certification program. As with any program that is contractual or officially represents your company, get the advice of your corporate legal team. Another good resource is Judith Hale's book on the subject, *Performance-Based Certification*.[1]

What Is Product Proficiency Certification?

Since product proficiency certification is a specific type of certification, one must first understand the definition of certification in general.

> cer·ti·fy
> \\'sər-tə-ˌfī\\
> Verb: to say officially that something or someone has met certain standards or requirements[2]

Product Training for the Technical Expert: The Art of Developing and Delivering Hands-On Learning,
First Edition. Daniel W. Bixby.
© 2018 John Wiley & Sons Ltd. Published 2018 by John Wiley & Sons Ltd.
Companion website: www.wiley.com/go/Bixby

Certification is the act of certifying, or having certified, someone. To certify someone is to say officially that he or she has met certain standards or requirements. Look at the important words in the definition of certification.

Officially. Using the word "certify" means that your company officially recognizes that word. Do not use it unless you have the permission and authority to define it properly.

Has met. You must have a way of knowing that individuals have met the requirements. Delivering a presentation rarely provides the opportunity to verify that someone has met the minimum requirements.

Certain requirements. You must define what the requirements are and they must be consistent.

As you can see, a certification course differs from a certificate course in several important areas. The most important one is that a certificate acknowledges the completion of a course or training process. Certification, on the other hand, signifies that one is able to perform a task at a minimum standard. Perhaps the most common certification is a driver's license. Most agencies require both a written exam (cognizance) and a driving exam (competence). When you pass both exams and the government awards you a driving certification, it does not mean that you are as experienced as other drivers are. It does mean that you have met the minimum standards to perform your new job: driving.

The term *product certification* by itself can be confusing. Due to increased globalization, most manufacturers are required to meet certain standards set by various industry and international organizations. These specification certifications control how companies manufacture certain products. They verify that a product does what it claims to do or that it meets minimum standards of operation. To differentiate between individual certification and merchandise certification, I prefer to use the term *product proficiency certification*, which is a certification that officially verifies that an individual can perform specific functions with a product.

In addition to the important words previously mentioned, product proficiency certification gets more precise with the general requirements.

Proficiency (to do). This certification means that the holder knows how *to do* something with your equipment. Make sure that action is well defined. It should be, if you created your objectives correctly.

Individual. Only individuals can be proficient in anything. This is one of the major distinctions between product certification and product proficiency certification.

When Do You Need a Certification Program?

Once one understands the meaning of the word certification, it is easier to determine what products warrant a certification program. The primary question one must ask is "Will our students benefit from a program that confirms that they know how *to do* something with our product?"

When Is a Certificate Program Sufficient?

Not all products need a certification program. It is much better to create a solid training program with measurable results than it is to risk the harm of an improperly managed certification program. Consider some of the benefits of a certificate-only program.

Faster, cheaper, and easier to implement. Certificate programs are generally faster and easier to implement, since they don't have the weighty requirements that certification programs do. Do not confuse this to mean they are less effective or easier to design or complete. A PhD is, in essence, a certificate program—and it is not easy to design, implement, or complete! Thankfully, most product training programs are not at that level.

It may meet all your requirements. It is very important to distinguish the difference between the need for an education program and the need for certification. The need for education on the product is not, by itself, an indicator that certification is needed. You may do well to consider offering an educational process first, until you can work out the full requirements of a certification program. In the process, you may find that a training certificate is sufficient. Even if you do not, the foundation of the certification program is often the preliminary education the students receive. Setting aside other factors and concentrating on creating a solid curriculum will improve your certification program when you do launch it.

Landing on the decision to create a certificate course instead of a certification program may be the best decision for you and your company. It is much easier to move from a certificate learning platform toward a certification program than it is to repeal a poorly or hastily designed business plan. If the education is needed quickly, get the learning out without sacrificing the future in so doing. Do not call it certification until it truly is.

Why You Should Consider a Certification Program

Now that we have eliminated the reasons for not choosing certification, look at a few of the many good reasons to choose certification.

If the Product Is Complex

While products are making life easier and more convenient, many are increasing in their complexity. More and more products manufactured today involve multiple fields of knowledge. The days of separating out mechanical products from electronic products are gone. Add in software and industry-specific technology and the complexities are compounded.

As products become more complicated, a comprehensive learning plan becomes more important. Experts can no longer rely on the general knowledge expectations of a particular field of study or industry. As products interlink more of the major branches of engineering, individuals who are outside of the formal education setting are gaining a broader knowledge of technology while losing the mastery of specific areas. Many technicians are continuing their learning in the workplace and focusing their learning on specific products—products that often incorporate multiple engineering disciplines. The learning developed for these individuals must take into consideration multiple backgrounds and a variety of competencies. Certifying that an individual is proficient in the product is often the only way to verify that they are capable and prepared to do their job.

A product can be simple to use, yet be technically sophisticated and complex. Making things easier to use is a big business in the technology world. However, ease of use does not come easily. A well-designed graphical user interface (GUI) took a longer time and was more difficult to develop than one that is more complicated to use. Another challenge is that the easier a product is to use, the faster that product changes. And the faster a product changes, the less loyal the customer is. Changes produce disloyalty because of a perceived lack of knowledge on the "new" product. Humans are most loyal to what they know or think they know. Product proficiency certifications are a great way to increase loyalty by building on past learning and tying what one knows from the past to the new product. Product proficiency certification can create a learning cycle that the certified individual will not want to break.

While products may be complex, certification programs do not need to be. Their main requirement is to verify that a student can demonstrate proficiency in the objectives of the certification course.

If Your Product Is Unique

Some products may be less complicated but are unique or new to the marketplace. They are often a good candidate for certification programs as well.

A product is not unique simply because it has unique features or specifications. However, if your product differs fundamentally from other competitive products, then it is unlikely anyone will be able to learn about your product from a different source.

Products That Are New to the Market

There are differences from products that are unique and those that are new. Products that are new—that incorporate new technologies or are otherwise at the front end of the product evolution phase—often change frequently. The main benefit of a certification program in these cases is that certification requires keeping current with the changes.

When the Go-to-Market Strategy Is Indirect or Complex

Just as products are more complex in their designs, they also go to market in a variety of business structures. The path a product takes from the manufacturer to end user can be complicated. It is often necessary to engage third parties to help us get our products to the marketplace. These third parties go by many different names, including value-added resellers, dealers, system integrators, distributors, manufacturing representatives, and others. Your industry may have its own name for the third parties you use, and the competencies they need in order to sell your products may vary.

Most resellers sell more than just your product. This should not worry you. A reseller's diversity of products represents an opportunity for trainers. People sell more of what they know better. If you can get your partners to understand your product better than they understand the competition's product, you will have an advantage. They will naturally lead with what is on the top of their mind, with the product in which they have the most proficiency.

Several years ago, I needed to buy an inexpensive, used car. I went to my friend, Doug, who owns a used car lot. I gave him my budget limits and asked him for advice. I had narrowed my options down to two cars on his lot that looked similar and had a comparable price tag. His response has helped to formulate my approach to customer training. He very quickly gave me the rundown on vehicle A. Much of what he said was neutral or even negative. It is true; there are a few honest used car salesmen out there! He gave me a list of things that could go wrong at different mileage points along the way. Normal things, things that one would expect on a used car. For example, I probably had another 60,000 miles before the transmission would need to be changed. The alternator, the timing belt, the gas pump, and even the A/C blower all had their scrutiny by my friend, who is as good a mechanic as he is a used car salesman. Then we moved on to vehicle B. Nothing. I pried for more information, but he had nothing more to give me. I was beginning to think there was nothing wrong with vehicle B and that I had found my next car when he completely surprised me by suggesting I buy vehicle A. "After everything you've told me, it's going to need done to it?" I asked. "Yes," he replied, "because I know that car like the back of my hand. I can work on it blindfolded. The other car could be great, but I don't know enough about it to sell it to you with confidence."

I have used this example dozens of times in seminars and effective training courses because it demonstrates an individual selling what they know better. If your go-to-market strategy is dependent on people who can choose between your product and a different one, training them effectively is not merely a matter of helping them identify opportunities or reducing their lead-to-sale time. It is likely a matter of whether they sell your product at all.

Of course, this applies to your own direct sales force as well. The principle remains the same. Even your own sales representatives will sell more if they understand your product better. There are many resources on direct sales training, and most companies understand that correlation with their own workforce. However, if it is true of employees who can only make money by selling your product, how much more true it is of a salesperson who could also make money selling a competitor's product.

Internal and external salespersons may require different training needs. A certificate program may suffice for one, while a certification program may better fit the needs of the other. A new-hire training program is often a certificate program. Employees who sell only one product line have the opportunity to improve their knowledge on a regular basis. Having a benchmark of what they are able to do (i.e., sell) is less important because companies have the ability to encourage that, lest they lose their privilege of employment. Sending them back to a certification program after several years of selling the same product may be unnecessary.

However, a certification program is often very effective with resellers. Manufacturers rarely have direct visibility to a reseller's sales tactics or internal talent development programs and may not even know their turnover ratio. A certification program offers a great opportunity to keep resellers up-to-date on your products. You cannot terminate another company's employee. However, as long as good certification rules and practices are used, you can determine that a third party employee no longer meets the requirements for certification. The goal, of course, is a positive one: to keep resellers more competent with your products than any other products in the marketplace.

If It Involves More Than One Party to Integrate

Many manufactured products are not just widgets sold directly to the end user. Most of the companies I have worked for manufacture complex products that require integration into other products, systems, or large infrastructures. Teams of people, ranging from high-level executives to blue-collar installers or tradesmen, may be required for successful implementation. In between are designers, architects, engineers, technicians, and others that require differing levels of knowledge on the product that is so important to us. Knowing that they have demonstrated an ability to complete their portion of the job has a myriad of benefits. Costs can decrease, while productivity, quality, and safety increase.

Whether the target audience of your certification program is pre-sales or post-sales, the value is in increased sales. Do not focus on the decision-makers and forget about the influencers. The more complex the product is, the more likely the decision to purchase was made in a cubicle, not an office. Someone wearing a suit may have signed the deal, but it was likely a technical expert who made—or at least strongly influenced—the decision. I have seen product sales increase from the thousands to the millions simply by educating the technical staff. All good managers and executives listen to their technical staff and allow their input on what will or will not work. When all else is equal, those experts will influence toward the product that they understand best. As an expert yourself, you have likely experienced this and are in the best position to influence this focus in your own company. Leaving experts out of the education loop can be detrimental to the future of any product.

If There Are Standards That Must Be Met

The Merriam-Webster Dictionary lists many definitions for the word *standard*. Addressed here is an accepted and expected requirement in your company or in a particular industry.

If There Are Industry or Company Standards That Must Be Met

Some broader industries expect or require individuals or companies to meet a defined set of guidelines. Experts refer to those guidelines as an industry standard. Of course, many government and municipal regulations are required for some products. I am not an expert on all of those and do not address those in this section. Referenced here are those requirements that either the general public or the industry leaders have determined are what "the good companies do." You do not want to be the company that does not live up to those standards.

Often, there are no industry standards to guide our training. A product or technology may be too new or so unique that the marketplace has yet to define any requirements. Some industries may be too secretive or even hostile toward the idea of collaborating on a standard. Even so, you or your company may create certain requirements or standards that sales or technical people are required to follow. This may include required experience in the field or outside education or certification. It may only consist of successfully completing a product training program or any combination of all of the previously mentioned.

When Quality Standards Must Be Verified

When the quality of the desired results requires control, a certification program may be helpful. It may be that your company wants to control certain tasks. It may be that your company wants to exceed the accepted industry standards. Certification may be beneficial in these circumstances.

As mentioned earlier, the way a product goes from production into the hands of the end user can involve multiple parties. If you or your company owns the product, those interdependencies are crucial cogs in a complex gear mechanism. If one party does not do their job well, it will affect the success of another party down the chain. Getting it right is not merely helpful, it is critical. If a product is incorrectly manufactured, it may mean that even the most qualified installer will not be able to install it properly. Likewise, if a well-manufactured device is not installed properly, even the best training may be futile in increasing its proper usage.

Take some time to write down all of the individuals that touch your product. What do those individuals need to be able to do in order to demonstrate proficiency in your product? If they do not execute that task well, what influence does it have on the individual next in line? If one's ability to perform is both critical and measurable, you may have a strong case for a product proficiency certification program.

If the Product or Technology Changes Regularly

One of the benefits of a certification program is the incorporation of a continuous learning cycle. While an individual generally only gets a degree or certificate once, a certification requires an individual to maintain their knowledge on a particular topic. This concept fits perfectly in a technology setting that changes frequently. Whether it is a new version of a software program, a redesigned machine, or an innovative approach to using an old technology, companies are always finding new ways to increase their influence or grow their business. As technology gets faster and smarter, it also gets more complex. A flashlight is no longer just a flashlight. It includes a GPS device, which is also a calendar, a game, a camera, an internet browser, and, oh yes, a phone. Even commercial farmers today are required to keep up to date on technologies that keep them profitable and sustainable.

Large software companies like Microsoft™ have done well at basing their certification on levels of change. If you need help with a current version of software, you will not be impressed with a technician who has a Windows 95 certification. If your product, or the technology your product uses, changes regularly, it may be beneficial to incorporate a certification program to ensure those tasked with creating, writing, designing, installing, configuring, or simply consuming are aware of, and proficient in, the latest changes.

If Misuse Could Cause a Safety Issue

Finally, consider whether any of those who touch your product on its way to market could easily misuse it. If a salesperson can easily sell your product to the wrong person or the installer creates havoc by installing it incorrectly, a proficiency certification program may help to ensure your product is used correctly. However, a certification program is not a *guarantee* of proper usage. Trainees may still misuse your product, either intentionally or accidentally. It is rarely a good idea to create a certification program out of fear, though it is necessary to be proactive. Product proficiency is a better offense than defense. The ultimate goal is still to increase sales.

The Requirements of Product Proficiency Certification

To have a legitimate certification program, there are several requirements. The requirements have been consolidated here into three groupings: proof of authenticity, proof of conformity, and proof of impartiality (Figure 18.1).

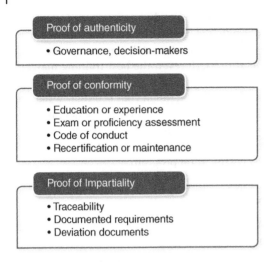

Figure 18.1 Three requirements for certification.

Proof of Authenticity

If you offer a certification to an individual, your company is officially recognizing that certification. Make sure that it is credible and authentic. Be certain that your certification does what it says it does and back it with the full support of your company or organization.

Board of Decision-Makers

A certification requires that you clearly define who has the authority to make changes to the program. In product proficiency programs, you should generally define this at the program level, not the departmental level. If your training department offers multiple certification programs, there may be different decision-makers for each program. Keep the list as short as possible but long enough to get the necessary input to make the program authentic. You may choose to include someone from product management, engineering, sales, or any area that includes stakeholders.

It is best if the instructor is not on the board, but some circumstances may render that impractical. The goal is to remove the responsibility of determining the criteria for certification from the instructor. The value in having a board of decision-makers is that it broadens the authority beyond one person or department and creates a backup for tough decisions and unforeseen deviations.

Curriculum and Program Acceptance

You should designate a diverse group of individuals to serve as approvers of the curriculum and the program. This may include all or most of the board members, but likely includes a few others as well. This one-time sign-off helps to generate agreement and endorsement from various individuals or departments in the company. Soliciting a marketing individual, a product management individual, a salesperson, and so on, helps to ensure that you don't exclude important details.

Proof of Conformity

Students must conform to the standards set as the minimum level required for certification.

Education or Experience

Most programs require some sort of educational process. This includes both prerequisites and any training particular to the certification. For example, you may require a formal degree, or a certain number of years of experience, as a prerequisite. Clearly define what that minimum requirement is. If the training associated is not a requirement, or you will substitute the training with a degree or experience, be certain that you have documented what is and what is not acceptable.

Exam and/or Proficiency Assessment

Most proficiency certifications require two assessments, one of a recipient's general knowledge and one of their demonstrated skills. Vet the questions well and seek additional help as needed. The completion of an exam, or at least a particular task, is usually required for certification.

Code of Conduct

In her book, *Performance-Based Certification*, Dr Hale points out that it is typical for corporations to include a code of conduct requirement in their certification programs.[3] This is helpful if your certification grants authorization to perform certain tasks. Your driver's license not only indicates an ability to drive but also grants permission to do so. The authorities can revoke your license if you violate the driving code of conduct. In the same way, you must be able to nullify a product proficiency certification if a student fails to live up to a minimum standard. Make the agreement clear to all participants. Retracting a certification should be rare, but you must have the right and process to do so. For example, if a certified individual takes a position with a competitor and uses their certification to harm your company, you should have the right to withdraw their certification. However, you will only be able to do that if they have signed a code of conduct.

Figure 18.2 shows what a code of conduct could look like. Your legal team must write one specifically for your company and circumstances.

Recertification or Maintenance

One of the main areas that distinguishes a certificate program from a certification one is the expiration of the certification. Most technical certifications expire after a set duration. That time varies depending on how quickly the awarding organization expects the technology to change. Some companies base the certifications on versions of the technology or software. Product proficiency certification can use either a time-based expiration or a version-based expiration.

Both a time-based and a version-based certification have expirations, so both require renewal. There are two common ways to renew or continue that certification.

Recertification. Recertification is a specific process students must go through in order to renew their current certification. Sometimes, recertification is as simple as retaking the class or exam. Other times, recertification may require a separate class and a different exam. Usually, companies or organizations grant recertification for the same duration as the original certification.

Some products may go for long periods without a change or have multiple versions in use at the same time. In these cases, companies may prefer a version-based certification program. Their "expiration" may simply be when a version becomes obsolete.

As a part of your certification process through [COMPANY], we ask that you carefully read and agree to adhere to the following code of conduct. This code of conduct is intended to ensure that all certified representatives of [COMPANY] uphold standards that reflect positively on themselves and [COMPANY].

As an individual with [COMPANY] certification, I agree to uphold the following principles:

1. **Competence:** I agree to maintain my skills and knowledge with relation to this certification and accurately represent abilities.
2. **Accountability:** I agree to take responsibility for the work that I perform and complete it to the best of my ability.
3. **Professionalism:** I agree to maintain a professional demeanor and quality of work at all times.
4. **Diligence:** I agree to work in a prompt, efficient and thorough manner at all times.
5. **Integrity:** I agree to perform work with honesty and integrity. I will distinguish between my personal opinions and those founded on qualified knowledge.
6. **Representation:**
   a. I agree to associate myself in all dealings with my legitimate company.
   b. I agree that I will not use or share the knowledge and materials I have gained to benefit [COMPANY]'s competitors.

Signature: _____

Date: _____

Figure 18.2 Sample code of conduct.

Maintenance. Maintenance certification requires that the holder fulfills certain requirements in order to continue extending their certification. These are typically in the form of continuing education credits, but one can adjust the requirements to fit specific needs. For example, one could require a certain number of projects to be completed or a minimum number of hours in the field in order to extend certification.

Instructor Certification Process

One important step often overlooked in a product proficiency certification program is the certification of the instructors. If anyone can teach a certification class, you may not have a true certification program.

The authority to designate this process belongs to the board of decision-makers. It is helpful, however, to offer a proven process for them to approve. The following is a sample process. You can use the steps in many product environments, though you may need to adjust the details to fit your needs.

- **Instructor product certification.** The instructor should first obtain certification by at least the same standards and requirements of any student.
- **Instructor training.** Since being a product expert and a training expert require two different skillsets, instructors should also go through training on how to teach proficiency effectively.
- **Teach or co-teach.** This process may require multiple iterations. Never send the instructor out to teach alone the first time. Using a coach or someone who is already certified strengthens the program and increases its credibility.

- **Sign-off.** Once the instructor is ready to teach on their own, designated approvers who have attended the class and observed the instructor should sign-off on his or her ability to instruct the class. Those will likely include someone who is already certified or otherwise known as a product expert and someone who is a training expert.
- **Maintenance.** Because this is a certification, it is usually best to require the instructors to teach the class a minimum amount of times in order to maintain their certification. They cannot improve their proficiency if they don't do it!

Proof of Impartiality

The third important requirement of product proficiency certification is proof of impartiality. This does not refer to being impartial to the products—the certification is specific (and, therefore, partial) to your products—but rather to being fair and making the rules and requirements the same for all customers.

Three major logistical efforts will help you to remain impartial and fair. Those three are the selection process and available opportunities of the program, the administration and traceability of the program, and the deviations and exceptions of the program. Proving impartiality requires thorough documentation and strict adherence to it.

Selection and Opportunity

If you plan to allow only a certain number of people through the program or otherwise plan to limit the opportunity to become certified, make certain that you document your decision process. Your selection process should be fair and legal. Always consult your legal department and get their approval that your selection process does not discriminate in any way.

Administration and Traceability

You must be able to track all students who are in a certification program. The tool you will use to do this will vary, depending on the number of students and products. Often, network security requirements force product trainers to use a different learning management system (LMS) than the corporate LMS, since many of the students may be customers, not employees. Regardless of the system you use, the key is to be able to determine an individual's status in the program quickly and correctly.

Track both the student information and the course revision information. You want to know what revision of the course the student went through, not just that they went through a course. This can be helpful when you teach a course that goes through multiple revisions.

Exceptions and Deviations

No matter how thoroughly you design your program, you will run into exceptions. No matter how much time you put into it, someone will always seek a legitimate deviation from the rules. You can be ready for it.

One of the jobs of the certification board is to review exceptions and deviations. This takes the responsibility away from the instructor and puts it in the hands of multiple individuals. The process should be simple and clear. Make sure you follow the process, even when the outcome is obvious. Document the decision in order to facilitate similar decisions in the future.

Documenting the Certification Program

Documentation is important for any training program. It helps maintain consistency and provides a framework for continuous improvement. Certification programs require documentation that is more extensive. Whenever the words "official" or "legal" are involved, it is important to keep excellent records and maintain clear processes and policies.

There are many ways to document a product proficiency certification program. Following is one way, but if you already have a documentation process that works, stick with it and be consistent. Your process should be easy enough to maintain but detailed enough to provide the necessary information. Create one complete document that outlines each certification program you offer. This document can reference any process or procedure documented separately.

Certification Program Document

The certification program document should include documentation of any policies that are specific to this program. If a policy is not universal but applies to more than one program, you may choose whether to include it.

- **Prerequisites and student selection and exclusion.** Clearly define who the program is for and who, if anyone, you will not allow into the certification program (i.e., competitors). Include the prerequisites for training and any related policies specific to this program, such as cancelation policies.
- **Course requirements and delivery.** State the main objectives of the course and the benefits of certification. You should also include the length of the course and any accreditation it may include.
- **Validation and continuation.** Document whether students are required to maintain their certification or renew it. Document what those requirements are and what the examination requirements are. This is also a good place to define who can sign the certificates and any other continuation or validation topics.
- **Code of conduct.** Define the code of conduct requirements. Since the code of conduct is usually short, it may be a good idea to include it in the program document even if the code of conduct doesn't change from one program to another.
- **Board of directors.** List who the individuals on the board of governance are and what the selection requirements are. Note their responsibilities in regard to the program.
- **Pricing and value.** List how you will determine the price for the class (if applicable) and what (if any) known exceptions you will make. In more mature programs, you may want to list what the budget or revenue expectations are as well.
- **Other specific documents.** All product certification programs are unique, as are companies and circumstances. There may be policies that are unique to your industry, country, or company. There may be documents regarding equipment or simulation processes. Add them to this list so that you have a complete checkoff of all documentation required. Delete any of the aforementioned that do not apply to your situation.

Process Documents

Separate from the individual program document are the documents that detail how to do the many functions related to each policy. Since these tasks usually do not change from certification to certification, document them separately. Most of them likely already exist.

Table 18.1 Certification documentation.

✓	Certification program (one document)	Things to document	Related processes (separate documents)	Document how-to	✓
○	Prerequisites and student selection and preclusion	• Prerequisites required for students • Target audience • Excluded audience • Cancelation policy	Registration, payment, and logistics	• Register for certification and/or training • Pay for the training/certification • Refund payments per policy • Provide proof of prerequisite fulfillment	○
○	Course requirements and delivery	• High-level objectives and expected outcomes • Minimum number of education hours and other course requirements	Curriculum revision and approval	• Vet and approve new curricula • Control revisions • Approve curriculum updates and revisions	○
○	Validation and continuation	• Recertification or maintenance requirements • Examination requirements • Certificate validation signatures, and so on	Instructor certification	• Certify a new instructor • Update instructors on changes to curriculum or program	○
			Reports, evaluations	• Process and deliver reports • Collect and report on evaluations	○
			Examination, certificates, badges, notifications, and so on	• Deliver exam in multiple ways • Deliver certificates and/or badges • Notify certification holders of pending expiration or changes • Student process for recertification or maintenance	○
○	Code of conduct	Code of conduct	Code of conduct violation	Report and address violations	○
○	Board of directors	• Names of board members • Board member requirements	Selection of board	Select board members	○
○	Pricing	• Pricing policy • Pricing exceptions • Budget or revenue expectations	Quotes, discounts	• Send a quote to customers • Process a discount request	○
○	Other				○
○	Other				○

Listed here are a few that are unique to certification. As with the policy documents, there are likely others unique to your situation. Use a checkoff list like the example list in Table 18.1 to verify that you have the necessary documents for your certification program.

Conclusion

Product proficiency certification programs are a great way to increase the loyalty, the quality, and the reputation of your product. If you choose to create such a program, make sure you design it correctly. Though your job may not be to create the program, you will likely assist those who do. Knowing what the requirements are can help you to avoid problems and promote a program that is defensible, profitable, and effective.

Making It Practical

Certification must be taken seriously. It is a powerful tool when used and documented correctly.

1) What three proofs are required for a certification program?

2) Describe the difference between a recertification and a maintenance certification program.

3) What process should be in place to certify the instructor of a certification course?

Before you read Chapter 19, "Managing the Details: The Effective Administration of Hands-on Learning," answer these two questions.

1) What three things does any company want to keep track of in regard to training?

2) In your own words, why is it important to follow up with trainees?

Notes

1 Hale, Judith. Performance-Based Certification: How to Design a Valid, Defensible, Cost-Effective Program. San Francisco, CA: Pfeiffer, 2012, a Wiley Imprint. This book is a very well-documented thesis on creating certification programs.
2 Merriam-Webster online dictionary. http://www.merriam-webster.com/dictionary/certifying (accessed August 8, 2017).
3 Hale, Judith. Performance-Based Certification, 2012, p. 58.

19

Managing the Details

The Effective Administration of Hands-On Learning

The administrative and logistical tasks required to run an effective hands-on training program may or may not be part of the duties of the product subject matter expert. Either way, two things are certain: there are many administrative tasks to perform and they are important to the instructor and all training stakeholders.

Technical trainers are often chosen for their technical skills. I've established in this book that learning how to teach is a distinct skill that product experts must choose to develop. In much the same way, training administrators are often chosen for their administration skills and forced to learn how to do their jobs as they perform them.

This chapter will not cover *how* to do the job of a training coordinator or administrator. It will help trainers, however, understand the importance of the administration role and provide clarity regarding its purpose. No training program is complete without the administrative function in place. There are three administrative functions all effective training departments must fulfill (Figure 19.1):

1) **Measurable.** The course and program must be evaluated for effectiveness using industry-based metrics and feedback.
2) **Sustainable.** The course and program must be repeatable with an established process for review, updates, or relevance.
3) **Traceable.** Records must be retained for reporting attendance, certification, compliance, and/or revenue.

The role itself is not the function of one person. Even if your company has an individual who has the title of training administrator, or training coordinator, the responsibilities belong to trainers and training managers as well.

Measurable

There are many reasons why effective administration is critical to the success of a product training program. Ensuring that a training program or course is properly evaluated is an important influence an administrator has on successful training. The reason the role is so influential is because it is the administrator's role to ensure a regular, systematic, and unbiased evaluation of the course and the program.

Product Training for the Technical Expert: The Art of Developing and Delivering Hands-On Learning,
First Edition. Daniel W. Bixby.
© 2018 John Wiley & Sons Ltd. Published 2018 by John Wiley & Sons Ltd.
Companion website: www.wiley.com/go/Bixby

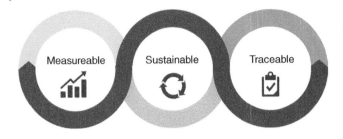

Figure 19.1 Three roles of training administration.

You Can't Improve on What You Haven't Measured

Said differently, you can't learn from failures you haven't noted. In a very real way, the administrator is the quality control director or continuous improvement manager of the training department. Working closely with the trainers, the administrator must keep track of what works and what doesn't.

Recently, my administrator notified me that we were getting weak evaluation results in one objective of a train-the-trainer class I teach. They weren't bad enough to raise any red flags, so I hadn't even noticed. She did. She observed that students had less confidence in performing that objective than others. After further scrutiny, I realized that some were misunderstanding one of the exercises I had created. I designed a new exercise and evaluated it separately for a few classes to ensure it was effective. It was, and what was a weakness soon became a strength. Had I not been asking the right questions and measuring the right things, I would have continued to assume I was doing a great job.

Chapter 7 covered some basic requirements for evaluating courses and assessing individuals. But all the assessment in the world is useless if it isn't acted on. Documenting what is being verified, piloted, or confirmed is critical.

Failure, of course, isn't a prerequisite to improvement. Good can become better. And it should. Just because you did something well doesn't mean you don't want to improve on your success. Find ways to ask the right questions so you can make every class just a little bit better than the last one.

But improvement isn't the only reason for measuring. Don't evaluate courses simply to find areas to improve. Track what you are doing well. It is just as important to measure what you are doing right as it is to measure what needs to improve. It is the continuation of what you are doing well that makes *measurement* bleed over into the important role of *sustainability*.

Sustainable

A great instructor teaching a great course today does not guarantee success tomorrow. Being a great training department is an iterative process that eventually leads to excellence. Excellence is at the heart of a learning organization. Experts and curriculum designers can invest hours into a learning course or module. It would be forgivable to consider the work finished after the first students have taken the class. Yet no one who has invested that much time wants the course to become irrelevant or outdated shortly after they launch the program. Worse, they don't want other instructors to teach the class they designed with any less care and determination than they would have put into it.

Controlling training content is important for practical and legal reasons. Content that is designed carefully should not be altered carelessly. It is impractical and a waste of time to have a different version of training on each instructor's computer. When the content is significantly altered, but offered as the same course, it has potential legal implications. It may become difficult to prove that all instructors are giving students the same opportunities or teaching the same objectives.

There are also financial reasons for ensuring the content is centrally controlled. Industry averages (and my own personal experience) indicate that it takes about forty hours of development for each hour of face-to-face training. When trainers alter that content on a whim—to satisfy a personal agenda or preference—it is a waste of time and money. Like the "telephone game" we used to play as children, the content soon gets altered beyond its original intentions and objectives. For these and other reasons all training programs must be repeatable, with an established process for review, updates, and relevance.

You Can't Repeat Success You Can't Define

Too many experts recurrently update their product training but don't track the changes or notify other trainers that they made them. Tracking the changes one makes to an instructor-led training course is important for consistency and control. Communicating those changes is important for sustainability.

Revision Control

Technical experts are very aware of the need for revision control. In fact, almost everyone today with a smartphone or other electronic devices is at least familiar with the concept of revision control in products. The same type of revision control is required for your curriculum.

Simple Revision Tracking

There are many ways you can choose to number your revisions. Product, language, region, topic, prerequisites, requirements, and even delivery method may be things you want to capture in your numbering (or lettering) system. If you only have a few courses, or are tracking courses only in your department or division, you may choose to keep it simple. Even a simple numbering or lettering system should track three things. It should track changes to the objectives, major content changes, and minor formatting or wording changes (Figure 19.2).

Figure 19.2 Simple revision tracking numbers.

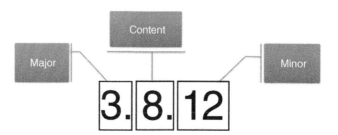

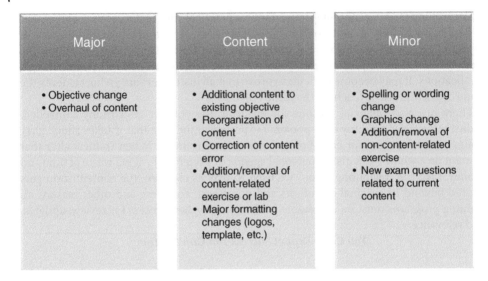

Figure 19.3 Simple revision tracking items.

One simple revision control uses a three-part numbering system. The first number indicates a major change. This should include any changes to the objectives of the course, or any major content changes that you want to record. The second number indicates changes to the content, if the objectives have not changed. It can also include major reformatting, reorganizing, or other significant changes to the curriculum. The third number indicates changes that are minor, though they may be important. This should change when even a slight change is made to the curriculum, including spelling errors and other small changes.

When a major change is incorporated, all of the minor and content changes should be incorporated into those changes. As such, whenever a change is made at the major level, the content and minor level numbers should be reset to zero. Likewise, when a change is made to the content level, the minor level number should be reset to zero, since all of those changes will be included in the content level change (Figure 19.3).

Having a numbering system is only the beginning. Keep a spreadsheet that lists the changes made to each revision. It is always a good idea to save the actual curricula for each content or major revision. Unless your industry requires it, you probably do not have to save a copy of the revisions with minor changes to them.

Global Enterprise Classification

If you are classifying and tracking the revision of a large, global program, you may want to use a more robust administrative metadata system. Some learning management systems (LMS) can help with revision tracking and even numbering, though most rely on an existing system. The goal of any data filing system is to simplify administration by making logging, retrieving, analysis, and revision easier, faster, and more practical.

If you use the 4×8 Proficiency Design Model for designing your courses, using a system like the following one will help with more than just the tracking. It will also help to identify and sort courses based on the criteria than can be tracked in each level (Figure 19.4).

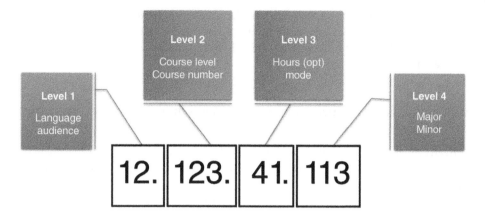

Figure 19.4 Enterprise classification system.

Level 1 (Optional)

The first level of the 4 × 8 Proficiency Design Model covers the business goal and the intended audience. Tracking at this level is optional, but, depending on the size of the organization, may be beneficial. Of the two steps in this first level, it is the target audience that can be designated in numeric form. Both the language and the target audience can be notated. Assign a number to each language you use or translate your courses into. That becomes the first digit (or two if you have more than nine languages) of the first set of numbers.

Some may prefer to use a two-letter designation for the language. The benefit of that is that it is easier spotted when looking for a particular course. The drawback is that it is not as easy to sort and adds more work if most of your courses are in one or two languages.

Both of these numbers are optional. You may choose to only record levels 2, 3, and 4, depending on the size of your organization or the scope of your classification system.

Level 2

Level 2 represents the objectives of the course. Throughout this book, I have emphasized the importance of clear and actionable objectives. Precise objectives are difficult to package into a filing system. However, when courses are organized or ranked based on their high-level requirements or other criteria, it makes it easier and practical. Hundred-level courses may be introductory courses, while 600-level courses may be reserved for train-the-trainer courses. The key is to define the meaning so that the numbers or letters you assign it are practical and valuable.

Level 3

Level 3 of the 4 × 8 Proficiency Design Model has the most steps. They include writing the outline, creating activities to help students learn, and determining how the learning will be delivered. Of these steps, the delivery medium is what should be classified. The simplest way to catalog the medium is to list the three or four ways your company delivers learning and assign it a single-digit number.

This is also the place to log the length of the class, if this is something you want to track. You can track either the actual hours of the course or the credits awarded, in whole numbers. Put this number prior to the delivery medium number, since it may be more than one digit.

Level 4

Level 4 consists of keeping track of the content. This is the simple tracking used earlier in a three-digit format at the end of the catalog number;

Hundreds level. The hundreds-level number refers to any major or objective changes. Note that if the overall objectives of the course change, the course should be given a new catalog number at *level 2*. However, if one of the objectives in the courses changes or is adjusted and you want to keep the course title and main purpose the same, this is where to record that change. All of the changes at the tens- and ones-digit level will be incorporated into any changes at this level. As such, whenever a change is made at the hundreds level, the following digits should be reset to zero.

Tens level. The tens-level number is for changes made to the content that do not change the objective of the course. If the curriculum is reorganized or something is removed, here is the place to note that change. All of the changes at the ones-digit level will be incorporated into any changes at this level. As such, whenever a change is made at the tens level, the ones-digit should be reset to zero.

Ones level. The ones-level number is for minor changes. Spelling errors, formatting changes, new exam questions, and many other less significant changes can be noted here.

Using the chart in Figure 19.5, which can be modified as needed for your program, see if you can determine the following from this catalog number: 12.123.41.113.

- What is the language of the course?
- Who is the general target audience?
- At what general level is this course offered?
- What is the course number?
- How many hours is the course?
- What is the course delivery method?
- How many major revisions has the course been through?
- How many content revisions has the course been through?
- How many times has it been updated for minor changes?

Using the graph in Figure 19.6, you can see the answers.

- What is the language of the course? **English**
- Who is the general target audience? **New hires**
- At what general level is this course offered? **Required**
- What is the course number? **23**
- How many hours is the course? **4 hours**
- What is the course delivery method? **Instructor-led**
- How many major revisions has the course been through? **One**
- How many content revisions has the course been through? **One**
- How many times has it been updated for minor changes? **Three**

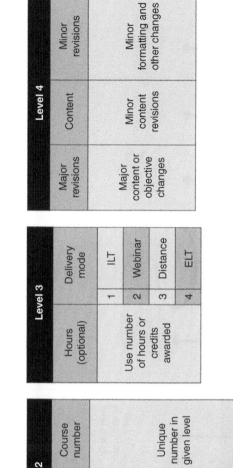

Level 1

	Language	Target audience
1	English	Internal
2	Spanish	New hire
3	French	External
4	Chinese	Contractor
5	Japanese	Sales
6	Arabic	Compliance
7	German	Regional
8	Russian	Business unit
9	Etc.	

Level 2

	Course level	Course number
1	Required	Unique number in given level
2	Prerequisite	
3	Prerequisite	
4	Prerequisite	
5	Certification	
6	Instructor	
7	Etc.	

Level 3

Hours (optional)	Delivery mode	
Use number of hours or credits awarded	1	ILT
	2	Webinar
	3	Distance
	4	ELT

Level 4

Major revisions	Content	Minor revisions
Major content or objective changes	Minor content revisions	Minor formatting and other changes

Figure 19.5 Sample enterprise classification items.

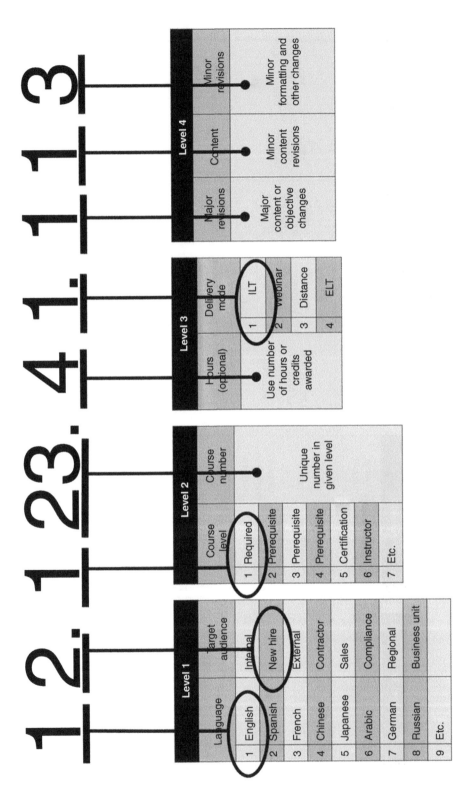

Figure 19.6 Example classification number.

Propose, Approve, Implement

All instructors of any course should have the freedom to propose changes. Each department, and maybe even each course, will have varying rules for approving those changes. Generally, the smaller the change, the fewer approvals will be required. Any instructor, for example, should be able to correct a spelling error or obvious formatting issue. They should be "approved" by notification only. However, changes that significantly alter the content or change the objectives should go through a review and approval process.

Making changes to your curriculum, no matter how wonderful they are, is of little help unless all instructors are notified of the changes. There are many collaboration tools available that can take the drama out of distributing a new curriculum to instructors. Find what works best for you and the size of your department or organization. If there are only a couple of trainers, personal communication may be sufficient. Larger, global organizations may require a more creative means of getting the word out.

The Quarterly Curriculum Review

Many sales departments perform a quarterly business review (QBR) with their employees. It is a chance to learn what is happening and what is important and make sure their numbers are on track. The goal is to give everyone the best opportunity for success.

The same can be helpful, on a smaller scale, with those who teach a particular curriculum. Companies that are serious about maintaining consistency throughout their technical education programs want to make sure that their instructors are on the same page, learn from each other, and have the latest and greatest updates to any curricula.

A quarterly curriculum review (QCR) can provide the best opportunity for technical trainers to succeed. This can be as simple as a conference call or can involve a full review and train-the-trainer session. No two QCRs are the same. Unlike QBRs, there may be many quarters without any changes, eliminating the need altogether.

A QCR is an opportunity to look at curriculum that hasn't been updated in a long time as well. This helps to avoid complacency and verify that your courses are relevant and up to date.

Change is difficult for everyone, even instructors. The advantage of having a QCR is that it allows instructors to voice their concerns or delights in the changes and helps to foster a team spirit. If the proposed changes came from outside the department, it also provides an opportunity to hear why those changes are important.

Oh, and by the way, this is a great time to share your success stories! Be honest, you smiled when you read that comment that you were the best instructor ever. Maybe the smile disappeared when you realized you might be the only person to ever see that comment. Now is your chance to make sure that doesn't happen. Share it! Encourage yourself and other instructors that the training being delivered is making a difference.

Train the Trainer

Taking time to train instructors on any changes to the curriculum is important. It is also important to create a process that requires verification that an instructor is ready to teach when the alterations to the course are beyond minor adjustments. Administrators should be able to track which instructors have been notified and, if necessary, been trained on the curriculum changes. Well-meaning instructors can unintentionally block sustainability by over-modifying their courses until they are unrecognizable from courses others teach. While there should be some room for personalization (I use the 80/20 rule—at least 80% of the course should be standard), most of that should come in the form of delivery, not in development of the curriculum itself. Developing a regular *keep-the-trainer-trained* program will help reduce too much individualization.

Prerequisites and Follow-Up

It is important for a trainer to be very involved in communications before a class and to follow-up with their students. Some of those responsibilities were covered in Chapter 13. The administrator, however, has the important task of documenting those conversations.

Prerequisites

The process of learning about your products is not started and completed in one learning transaction. That does not minimize the importance of the training event. It is important to get as much out of the education process as possible. The more prepared students are before they come, the more they will learn from the class. The more distractions an administrator can help eliminate, the more focused the students will be in class.

Here are a few things to consider as part of the pre-communications. These are in addition to the list in Chapter 13. Add your own items to this list as necessary.

- Consider creating a welcome guide. It can include things like...
  - Directions to the training event
  - Local hotels
  - Local restaurants
  - Local shopping: malls, pharmacies, and so on.
  - Tourist sites
  - Current weather information
  - Public transportation options
- Send information about the instructor. This is often better done by a third party. "She is an expert in..." sounds so much better than "I am an expert in...."
- Send any pre-reading material between 1 and 2 weeks ahead of the class. This should include the code of conduct, especially if it is a certification course.

Follow-Up

Don't stop the communication after the students leave. Always continue the conversation after the classroom event is over. Many administrators or instructors send follow-up emails to their students for one of two good reasons.

Many instructors communicate with their students after a training session because they feel it is the polite thing to do. They want to show their interest in the students and make themselves available for questions. That is honorable and important. Others follow up students in order to get feedback. They may send a survey or further evaluation with the goal of improving the class in the future. That, also, is an honorable thing to do.

Those are both good and important things to do, but they are not the main reason why continuing the conversation with your students is important. Following up with students is not just a polite thing to do and it isn't just about getting them to help us do better in the future. Follow-up has an important role in the success of the training that was just delivered. We have all read discouraging statistics about how much people remember after a training event. Hands-on learning will help to increase what people remember, but it isn't a cure-all. Reminding students about what they have learned and how they can apply their learning on the job will help to make the learning more successful. When the learning is successful, you will be successful.

Here are a few things to consider as a follow-up after the class. These are in addition to the instructor list in Chapter 13. Add your own as necessary and appropriate.

- ❏ Invite students to any existing knowledge group where they can continue to share and learn.
- ❏ Remind students of their commitments and learning.
- ❏ Send a delayed survey for their feedback on the class after some time has gone by.
- ❏ Remind them of further learning opportunities.
- ❏ Ask for key successes and how they have applied the learning.
- ❏ Ask how you can better prompt them of their learning when they are on the job.
- ❏ Ask them to recommend the class to other students or their leaders.

Good follow-up is the right thing to do for the student. If you can trigger their memories enough to apply what they learned in the class, you will be helping them. When you help them, you will be creating a better training program.

Traceable

The third administrative role of any training department is to ensure that you know who has taken what classes and who taught them. This is important for professional, legal, and practical reasons. Many companies are finding that much more training is being delivered than they are aware of. This can create problems when a customer claims to have gone through "training" from our company, but doesn't show up on any records.

You Can't Be Rewarded for Success You Can't Prove

Success isn't all about numbers. But successful training departments must know the numbers. You need to track people and programs as well as business results. The statement that you can't be rewarded for success you can't prove has a related, much more sinister, and dangerous alternative statement: you can be held liable for training you can't prove.

Protect yourself and your company. Certificates with your company logo on them could be used as legal evidence of how or what a student was taught. Never allow employees to hand out certificates with the company logo on them unless that information is being tracked. If you have any questions about this, seek advice from your legal experts.

Tracking People and Programs

Most training departments will use an LMS to help them track their students. If you can use your corporate LMS, you may choose to do so. Many smaller technical training departments find that they cannot use it, either because it is accessible only to the Human

Resources department or because it is behind the company firewall and they need access for both customers and employees.

This book will not go into the details of choosing a good LMS. There are many, many available. They range from open-source structures to expensive enterprise systems that include proprietary content. They all have their specialties and some are better at certain things than others. If you are in the position of considering an LMS for your hands-on product training, there are many things to consider. Create a spreadsheet and compare each LMS based on these and any additional criteria you have. For ideas, download the comparison sheets from LMS websites. Use multiple ones, since a provider will not ask you to compare something they are not good at!

Always get a demo before you purchase. Most providers will allow you a free trial period and I suggest you take advantage of the offer. You can learn a lot when you try to apply it in your own setting and under your circumstances.

There are a couple of key things that product trainers need to consider that are not available in all LMS programs. They may or may not apply to your training.

1) Does it manage individuals only, or can it manage companies?
   All good LMS help track an individual from point A (no training) to point B (fully trained). Fewer LMS are capable of tracking an entire organization's progress.

   Suppose you are running a program that requires a company to have five trained (or certified) employees on your product in order to maintain their authorization to sell or install your product. Can the LMS manage that?
2) How does the LMS handle classroom training?
   Almost all LMS will *say* they can handle classroom training. Most are built first for eLearning. That's easy, since it is an electronic management system managing electronic learning. It is harder to create a good LMS that handles classroom learning in a way that doesn't complicate life for the instructor. Know what you are getting into before you buy.

   a) How do students register?
   b) Is the LMS able to handle external registrations? Does it have an external marketing/registration page?
   c) Can one person register several people from their company?
   d) How does the system handle no-shows and cancelations?
   e) How do you use it to track prerequisites and follow-up work?
   f) How does it handle private classes?
   g) How flexible is the database? Can the data be exported to other programs? Caution: Do not use an LMS that does not give you complete ownership of your database.

3) How does it handle payment for classes?
   This is important if you are charging customers for your product training. If you have a tiered go-to-market system, this may create further questions.

   a) If it requires my corporate merchant account (unless you own your company, you may have to jump through many hoops to get this), will they help with that process?
   b) Does it allow for multiple payment methods?
   c) Does it have a discount method (vouchers, discount codes, or other means)?

4) How do evaluations and assessments work on the LMS?
   Too many instructors are forced to bypass the LMS because it isn't practical as an exam delivery tool. You should at least know what you are getting.

   a) What types of questions can it handle?
   b) Does it allow for question pooling by topic or objective?
   c) How is the exam administered (online, on paper, etc.)?
   d) Does it print or email (or both) the certificate?

5) What does it look like to the end user?.
   If you are using this for customer-facing training, the aesthetics of the LMS are important. It represents your company and the quality of your training program.

   a) How many clicks does it take the customer to register for training?
   b) Does the look and feel represent the message we want to deliver?
   c) Does it share the information I want and only the information I want to the customers I want to share it to?

   Of course, there are many other considerations that are not unique to classroom or customer training. They are important to consider as well.

6) Is it cloud-based or managed internally?
7) How easy is it to learn and use for the administrator?
8) What other types of delivery platforms can it handle (i.e., mobile)?
9) What types of reporting can it handle? How about customizing reports?
10) Finally, how much does it cost?

   a) Almost all systems require a start-up cost. Find out what is included in those costs in detail.
   b) Do they charge by student, by course, by flat fee, or others?

There are many different ways to charge for an LMS. Most providers will offer more than one way to pay, since every company and every need is different.

Tracking Business Results

Tracking business results is a much more difficult task than merely keeping track of who has taken a course. The responsibility of tracking business results often falls on a training manager, but it is impossible to track without the help of instructors, coordinators, and (almost always) a number of people outside of the training department.

   Business results are most often tested as a return on investment. Simply speaking, corporate leaders want to know if the money they are allocating to training is a cost or an investment. More importantly, they want to know that the training is helping the organization meet a greater strategy. The only way to know that is to track what is important to the company. In their book *What Makes a Great Training Organization?*, Harward and Taylor state that being aligned strategically "is the single most important factor that impacts whether a training organization is viewed as... being great."[1] The challenge is that your product training likely falls into one of four business reasons. To complicate it further, if you teach more than one product training class, you probably have some classes in each category. Each needs to be tracked for its own reasons.

Tracking Compliance

Depending on your industry, there are multiple types of compliance training. I'm including here any training that is required, either by law or by the general practice of your industry. Tracking compliance training seems obvious. Most agencies won't just take you at your word that John Doe took the training. Most have defined items to track and I will not attempt to even begin listing them. If you lead, teach, or administrate compliance training, you are much more aware of your industry requirements than I am.

Tracking Revenue Generation

This book is not about how to run training as a business. David Vance and others have written large volumes on this topic.[2] In Chapter 5 I make a case that there is a difference between running training as a profit center and generating true revenue.

The best way to track revenue is to track the sales of individuals or companies that have gone through training and compare it with those who have not. You will almost always need a control group—a couple of companies who have not had the training. They may or may not know that they are the control group – that depends on your circumstances. The challenge will always be to eliminate other factors that could alter the results, either positively or negatively. I suggest running a 6-month control group every five years or whenever you create a new, revenue-generating class.

Tracking Cost Savings

Training that saves costs is slightly easier to calculate, but equally important. Here also, it will depend on the type of training. You may be able to calculate reduced technical support calls, installation time, up (or down) time, and so on. Run these analyses as a study; don't try to run them constantly. There are always factors that influence the data that are beyond your control. If you run it for a shorter period of time, you will more likely be able to manually take out those factors. If you run them for long periods of time, your data may be discredited by one or two events beyond your control.

With both revenue-generating training and cost-saving training, you should be looking for opportunities to make these assessments at any time. Solicit the help of technical support leaders, sales leaders, and others. I once had a technical support manager approach me with an opportunity to compare two companies who were embarking on a very similar project. One had invested heavily in training, and the other had not. We followed them through the 3-month project and found that our return on investment was several hundred times better than what we had been advertising! Of course, not all results are that outstanding, but the ability to measure it and the partnership with other leaders created an opportunity we might otherwise have missed.

Improve Services

This training may be seen as goodwill training. It is the training your company provides just to save face or to make friends in the industry. This is the training that salespeople ask for with a statement like "We need to give them this training because they are the leader in the industry," or something similar. The truth is that most training that experts put into this category could actually fall into the revenue-generating category if anyone knew how or took the time to measure it. If possible take that time and find that way. This type of training is fun to do, but hangs by a tenuous thread. When times get tough, goodwill tends to disappear!

Conclusion

Effective product training must be effectively administrated. Many years ago, I worked for a CEO who stated that the receptionist should be the highest paid person in the company. I don't know if he actually did that (I never asked her!), but his point has always stayed with me. No matter how good your product is, if your customers get a bad first impression, they will be hard to win back.

The same is true with training. Great training can be ruined by bad administration. Good administration can help to cover a multitude of faults. It is important, not just because it makes training look better. It actually makes training better. Hands-on learning works because it builds on a student's own experiences. Good administration works when it helps the whole department build on its own collective experiences.

Never underestimate the power of good training administration. You must measure your success in order to be rewarded for them with an even greater training program.

Making It Practical

The administration of a technical training program may seem like a simple or unimportant task. The details of training have implications beyond organization and functionality. Effective administration actually improves the learning possibilities.

1) In your own words, how does the administration of training affect the learning outcomes?

2) What are the three things that every learning department must administrate effectively?

3) What are the three things that must, at minimum, be recorded in any tracking system?

Before you read Chapter 20, "Developing New Product Talent: Effective Mentoring of New and Junior Employees," answer these two questions.

1) What have you learned from observing other professionals?

2) In your own words, why do you think mentoring matters to companies?

Notes

1 Harward, Doug, Ken Taylor, and Russ Hall. What Makes a Great Training Organization?: A Handbook of Best Practices. Upper Saddle River, NJ: Pearson Education, 2014, p. 19.
2 Vance, David L. The Business of Learning: How to Manage Corporate Training to Improve Your Bottom Line. Windsor, CO: Poudre River Press, 2010.

20

Developing New Product Talent

Effective Mentoring of New and Junior Employees

This book is primarily written for technical subject matter experts. Think about that term for a little bit. Subject matter expert. Perhaps you feel it is overused. Maybe you wear the title with pride. Either way, you likely are one and the designation comes with responsibility. You worked hard to get the technical knowledge and skills you possess, but you did not get them alone. Even the greatest inventors only discovered a new or better way to do something. They did not create something out of nothing. Now it is your turn to help others. Perhaps that will come in a formal training setting, but it may also happen through mentoring a new or more junior employee. This is not a task to take lightly.

The need for technical experts to share their knowledge is greater today than it ever has been. There are many books and articles written that highlight and even dramatize the pending doom if the aging workforce doesn't more adequately educate the new workforce generation. Statisticians track the number of engineering and technical diploma graduates as a way of measuring the technical health of a society. But those in charge of the technical knowledge of a company have the added concern of something harder to quantify: experience.

Why Mentoring Matters

Most people like the concept of mentoring. Even if they don't, most would fear some sort of social backlash if they vocalized their disdain for being a thoughtful and caring human. What fewer have taken the time to consider is *why* mentoring matters. The truth is, unless you are specifically charged with developing new talent, your concern is less with *why* you should educate new talent and more with *how* can you do it. Beyond the aging workforce argument, however, there are at least two benefits (or *whys*) to consider that will directly affect *how* you transfer your knowledge to other internal employees.

Product Training for the Technical Expert: The Art of Developing and Delivering Hands-On Learning, First Edition. Daniel W. Bixby.
Companion website: www.wiley.com/go/Bixby

Why It Matters to the Mentor

The first thing to consider is why it matters to you, the potential mentor. There are two good reasons to become a mentoring expert. The first is because experts who mentor other employees are more valuable to their organizations. The second is because mentoring experts are more fulfilled in their roles and in the legacy they will leave behind.

Employers Value Mentoring Experts

Experts who can help others learn more are very valuable to technology companies. As computers improve our ability to forecast and companies learn from past mistakes, the gem of a teaching or mentoring expert is becoming more obvious. Companies value teamwork and some even reward their engineering teams based on their ability to improve as a team, not as individuals. There are four main reasons why this is truer now than ever:

There is more technology. This may seem obvious. The rate at which technology expands is difficult to process. Universities and technical colleges can't keep up with the regular changes. Many have resorted to hiring professors who are experts in their fields while giving little or no care about the quality of education those professors are able to impart.

Technologies are changing faster. This is related to the statement above and is part of the reason why there is more technology available today than ever before. However, this relates specifically to existing technologies that become obsolete due to changes. These changes may be made in attempts to catch up with the competition or in efforts to stay ahead of them. Either way, change happens and it happens fast.

Products and technologies are more proprietary. Most companies plan on investing a significant amount of time and training in new technical hires. While the universities provide the basics and the foundation for their learning, even the most experienced hires have to learn proprietary technologies and methodologies—and there are a lot more of them. As products become more complex, so do the number of patents that make them unique.

Some things are better caught than taught. Technical differences are not relegated to things that can be patented. Material designs, applications, manufacturing techniques, software differences, and any number of innovative differences may be insignificant when compared with the culture or philosophies of a company. Sometimes those things are learned best when a new employee, working alongside a senior technician, sees them put into action on a daily basis.

Of course, each of those reasons has a potentially negative downside:

More technology to learn means that companies take a greater risk when hiring new employees. Degrees or even past experience does not always mean an aptitude to learn a different technology.

Changing technologies means that experience more quickly becomes outdated.

Proprietary technology makes employees harder to retain when the competition wants their knowledge.

The fact that **some things are better caught than taught** can be downright frightening! I'm sure you can relate to the following real example:

No matter how much safety training a company requires, if senior employees take safety lightly, their unspoken message will override any formal training.

When you consider the positive reasons for mentoring, it seems that these four things combined would compel any ambitious company to require their senior experts to mentor new employees. When you consider the negative reasons, you understand why many do not!

The truth is that mentoring is most helpful when it is seen as a partnership. Senior employees need to learn from new employees almost as much as the vice versa. What they learn may be very different, but the healthiest companies create opportunities for the learning to happen in both directions.

Successful Experts Are Teaching Experts

Success is more than just completing a task or obtaining financial security. True success includes a sense of fulfillment and investment in what matters most—people. By this definition, the most successful experts are those who freely share the knowledge they've been lucky enough to gain. Their success comes because they are trusted, both by the company and by their fellow employees. Said differently, trusted experts are teaching experts.

Of course, there are experts who will not mentor or share their knowledge. They hoard information as if they have done something superhuman to deserve and own it. They may do fine in their job. They may make a decent living. They may even force the company to pay them more for the knowledge they won't share. But one thing they will never be is trusted.

Trusted experts are more likely to survive product failures and economic downturns. In fact, they are the ones companies turn to as an advisor when difficulties arrive. Good salespeople know that to be successful they must be able to gain the trust of their clients. While their focus is on external customers, the same is true of internal relationships. Only those who are willing to share the knowledge they have gained that has contributed to their success and that of the business will be able to influence other people. If you want to leave behind a legacy, it is best to leave it in people, not a product.

Why It Matters to Your Company

Mentoring is not just a benevolence that companies use to demonstrate their support for mankind. It isn't a "good neighbor" program they are willing to lose money on in order to promote their brand or image. It really matters. It really makes a difference.

Mentored Employees Have Real Input Sooner

One of the goals of talent development specialists is to help employees take ownership and satisfaction in the value of their given job. Everyone wants their job to be meaningful. Employees who feel like their input is minimal or nonexistent are less likely to get engaged. One of the fastest ways to get their input is through a mentoring relationship.

Allowing a new or junior employee to work alongside an experienced expert to accomplish tasks that they otherwise would not be able to undertake is one of the quickest ways to advance an employee from entry level to peak performance. The tasks must be well defined and appropriate, but the experience is invaluable.

More Meaningful Experience Sooner

Mentored employees have the opportunity to gain more experience sooner and that experience will be more meaningful. In a well-designed mentoring program, all employees have at least four distinct types of tasks they could face on the job:

Accomplish: Tasks they are responsible to complete. All employees, even those on their first day of the job, have certain tasks they are responsible to complete. These tasks grow over time and vary from day to day, but they make up the bulk of a typical employee's day. I refer to these as level 1 tasks, since they are generally what the employee was hired to do.

Augment: Tasks they are seeking to develop. All employees should have at least some tasks or skills that they are seeking to improve or develop. These are what I refer to as level 2 tasks. This may include a few tasks that are part of their daily requirements but should also include some tasks beyond their immediate responsibility. These are duties that are within reach and do not require a significant amount of extra experience to perform. Level 2 tasks can also refer to responsibilities they *could* perform but don't, either because it is not part of their job or because they would simply need more information to complete it.

Assist: Tasks they can help an expert to complete. This is the heart of a mentorship program. When all employees have at least one task they are working with someone else to complete, the learning will multiply. Similar to a stretch goal, these tasks are beyond the everyday reach of the employee but within their capability to provide meaningful assistance. These are level 3 tasks.

Advise: Tasks they can teach others to do. Even new employees can benefit from the opportunity to coach others. Everyone is an expert in something. Finding that opportunity and allowing them to mentor others not only makes the employee feel valuable, but it also helps to support the philosophy that mentoring is for everyone. These are level 4 tasks.

Mentoring for Proficiency

Not all mentoring is equal. In order to get the most out of it, at least in regard to product proficiency, a mentoring program must be intentionally designed. Three emphases will help to make product mentoring more effective. First, use as many mentors as feasible. Second, make the experience real and valuable. Third, make the mentoring a partnership, in which both parties are expected to take away some learning.

Multiple Mentors

Some good mentorship programs pair one employee up with another for a long period of time. These programs work well when the intent is broad in perspective and deep in potential. It works well for transferring soft or undefined skills. With product training, however, it can often be more helpful to observe and learn from multiple mentors. These activities are not a training class; they are an opportunity to make a difference by supporting multiple people. Different people will need different types of assistance, and the more exposure one gets the better.

Real-Time Mentoring

The emphasis here is on the word *real*. No matter how much effort we put into a training class to make it seem real, it isn't. But at some point, the tasks they execute on your product will be very real, indeed. No one wants a doctor to perform surgery who hasn't assisted in the same surgery multiple times. The idea is to blur the line between "first times." Everyone has a first time. They should have two—the time they first observed another professional doing it and the time they first did it themselves. Those events may be well defined and specified legally or otherwise. Experientially, however, that is usually not the case. The multiple experiences of performing that task with another professional should make it difficult to define when the "second" first time happened.

Partnership Mentoring

All those involved in the program can learn from it. No one is exempt. In order to make this a partnership, both sides need to grow from the experience. Experienced experts can learn from future experts. Young talent is more likely to trust mature talent when they are treated with respect. One of the best ways for a senior employee to show respect for a new employee is to admit, and even detail, what they have learned from them. When a senior expert shares what he or she has learned with other team members, there is almost always a boost in trust and respect, two things that help keep a team together. When teams trust and respect each other, they will not hoard information or knowledge from their teammates.

The Foundation of a Mentoring Program

Formal classroom teaching isn't the best way for every expert to transfer their knowledge to their peers. As such, it is important to provide a way for everyone to become a teaching expert and gain the trust of their company and colleagues, even if the teaching is informal and on the job. Informal does not mean unstructured. The companies that profit the most from mentoring are companies that provide some structure for doing so to their employees.

Develop a Structure for Success

Like any good program, your mentoring efforts require a good plan. A corporate-wide plan should probably be initiated, led, and monitored by the talent development organization. If you are preparing to lead a corporate-wide mentoring strategy or even a program that expands beyond your immediate team, ask your human resources team for assistance. They may have someone who can help lead a program like this and relieve you or your department of the administration work that may be involved. This is especially important if involvement will affect an employee's required development or go on their permanent records.

However, some technical groups or individual managers may want to create a plan for their smaller group. The following are some things to consider if you want to help other experts share the product knowledge and experience they have with their peers.

Get Appropriate Endorsement and Approvals

Some individuals may choose to incorporate mentoring into their individual development as well. Any plan may require approval from the supervisors of both the mentor and the mentoree. But authorization to proceed will not guarantee success. You need more than that. You need to get the endorsement and support from senior management in your organization in order for your program to have continued success.

Before you seek the support of management, get as much input as possible regarding how much time you can and should allocate to both mentoring and being mentored. List the experts you or your team can learn from and then solicit their help. List specific tasks that can be completed by a mentoree. Find examples that provide the exposure you want. The more organized and deliberate the information you provide is, the more likely you are to get the full backing of management. The more input and feedback program designers get, the more effective the program will be.

Set Realistic Goals

Define what success looks like for the program in general. Start simple. Set realistic expectations for how much time must be invested by both parties. Document how you will measure success and how often you will reevaluate the program. Don't force mentoring for the sake of having a program. Start with a few champions and, as much as possible, let the program grow voluntarily.

The goal is to create regular interactions that are long enough to be meaningful and short enough that they don't significantly interfere with either employee's level 1 tasks. New employees may have higher mentoree goals than veteran employees. Even more senior-level employees can benefit from a program like this, though you may choose to call it something different to eliminate seniority or hierarchy concerns.

Create Individual Objectives

There are generally two different approaches to creating mentoring objectives. One is to seek out specific tasks to learn and assign partnerships accordingly. The other is to allow the tasks to come naturally in a job-shadowing environment. Both are important. Either way, each mentoring partnership should have some clearly defined objectives for the individuals involved.

When the objective is specific, pair less qualified individuals with more experienced individuals who will be involved in that particular competency in a real environment. Define the product experience the new talent should obtain. List the types of experience or knowledge you want the senior talent to demonstrate. Ask the mentor to act as a guide to develop the new talent. They should also provide feedback on progress and be the final word on whether the new employee can accomplish the task.

When the objective is to gain broader knowledge of the industry, a customer, or your own company, it may be appropriate to allow a new employee to shadow an experienced employee in their daily job. This is also a good technique when the tasks cannot be dictated, as in, for example, a technical support environment. In these cases, the objective may be an amount of time, or until a senior expert deems the learning to be sufficient.

Define the Qualifications of a Good Mentor

Just as it is never wise to ask an expert to teach a class without helping them become a better instructor, so it is unwise to expect anyone to be a good mentor without first defining what that means for them. There are two things that are required for anyone to be a successful mentor. The first is aptitude and the second is attitude.

Aptitude

The fact that any person becoming a mentor must be proficient at the task they are expected to teach may seem obvious. Too often it is assumed. Never pair a junior employee with a senior one without first taking into consideration the specific skills that need to be transferred. Not all senior employees are good at everything, and not all junior ones need to improve everything. All employees have something they are adept at. The challenge may be to determine what it is.

Aptitude, however, goes beyond the technical, product, or soft skill abilities of an individual. Good mentors must also be good at mentoring. In order to become a good mentor, subject matter experts need to have their role clearly defined. Product experts must understand that showing a mentoree how to do something is not enough. Effective mentoring is similar to coaching. It becomes a two-way dialog. Mentors must encourage and provide candid but constructive feedback in the same way a sports coach might help an athlete perfect a particular skill. To do so on the job with your product, expert mentors must allow the learner to perform specific tasks. In this way, the learning will become real and practical.

They also need to learn the tendencies experts have when conveying technical facts. They need to understand that they likely operate in the fourth level of competency and will need to remember what it was like to learn the material and go back to the third level of competency in order to teach. Refer to Chapter 3 for more information.

Mentors should also understand how the four principles of learner-readiness apply to mentoring (see Chapter 4). If necessary, the mentor must help the new employee understand the value of the learning they are going to get (recognize the need). They must help the student take ownership and responsibility of their learning. A good mentor will relate the present learning to any experiences their mentoree may have had, and they will look for opportunities for the pupil to apply it as soon as possible.

A program leader could use the first part of this book to help mentors get a better understanding of how adults learn. Whatever you use, set the expectations and give the appropriate help to ensure success.

Attitude

It is almost always a bad idea to force a subject matter expert into becoming a mentor. Most successful programs are volunteer programs. If the desire to be a mentor isn't there, the likelihood of success is minimal.

Lack of desire may be due to several reasons. Some employees may feel threatened by such a program. Others may have had negative experiences that have created a bias against mentoring. Still others may simply see it as a waste of time, forgetting the importance of experience in their own learning history.

Because these adverse attitudes may exist, it is important to address them before launching a program. If necessary, create incentives and reward good examples. Remove obstacles and reduce the workload of potential mentors, so that mentoring is not an extra burden they are not prepared to bear. Employees will change their attitudes if they understand the true value of the program for the mentor and if it is feasible for them to do so without feeling overworked.

Conclusion

Mentoring programs work because they allow adults to learn in the way that is the most natural and the most effective—by doing. It gives new employees an opportunity to slowly and informally absorb learning from a more experienced expert, one who may or may not be a great classroom teacher. It provides an excellent opportunity for knowledge to become practice and for lab experience to become real-world experience.

To make the most of a mentoring program, get as many experts involved as possible. Pair new employees up with as many experienced employees as feasible. Provide the structure and support required for success and take notes of what works and what doesn't. No two mentoring programs will look exactly alike. Industries, technologies, products, cultures (both corporate and regional), and many other factors will alter the precise way you use mentoring as a program. What doesn't change is the value of allowing experience to create more experience. When that happens, it won't be long before the new employee is the one doing the mentoring.

Making It Practical

Mentoring can be rewarding to a seasoned expert and be valuable to a company, especially a technology company.

1) Why do you want, or not want, to be a mentor? What about being mentored?

2) What are some things you feel are important to consider when designing the structure of a mentoring program?

3) What is one of the main reasons mentoring is successful as a teaching tool?

Review of Part Five: The Operation of Hands-On Learning

1) What are some things your company does well in regard to the operational functions of hands-on learning? What are some areas you could improve on?

2) What are some things you are going to do differently as a result of reading Chapters 18–20?

21

Now, Go Do It

To Be an Effective Trainer, You Must Train

Every good training class, every good presentation, and every good book has a call to action. This is your call to action. The principles in this book apply globally. Whether you are teaching in your hometown to local students or halfway around the globe in a foreign language, these principles will help make you an effective trainer. But they will only help if you go do it. To be an effective trainer, you must be a trainer. Just like your students, who must install your product in order to be an effective installer, or sell your product in order to be an effective seller, you must teach in order to be an effective instructor.

Two things make a great instructor: a love for learning and a love for people. Notice that I did not say a love for teaching is required. That is secondary. I hope that a love for teaching will come, but it is not required in order to be a good instructor.

When you combine a love for learning with a love for people, the result is a love for seeing others learn. When that learning makes a difference, you have become an effective instructor.

This book is just a step in the right direction. Create a plan that will increase your effectiveness as a trainer and keep you improving. Define your approach develop with a strategy, design with a structure, and then go out and deliver with a purpose.

Define Your Approach

DO Articulate How You Will Make Learning Effective

Can you articulate the philosophy of learning that guides all of the training you do? Can you state how and why your training works? The first step toward being a consistently effective instructor is being able to articulate why you are effective.

DO Emphasize Proficiency over Knowledge

Your documented approach does not have to be long or deep. It merely needs to reflect that your students will learn by doing—by putting into practice and demonstrating the objectives you want them to complete. Knowledge is important, and you can make it

Product Training for the Technical Expert: The Art of Developing and Delivering Hands-On Learning,
First Edition. Daniel W. Bixby.
© 2018 John Wiley & Sons Ltd. Published 2018 by John Wiley & Sons Ltd.
Companion website: www.wiley.com/go/Bixby

even more important by making it actionable. Only when students turn knowledge into action will the knowledge become valuable to them and to your company.

DO Become Consciously Skilled on Your Products

The "secret sauce" that makes product experts that can teach others so valuable to a company is that they know what they know about their products and know why they know them. Being able to perform tasks with your product without even thinking about it makes you a great employee and subject matter expert. Being able to perform those same tasks and explain everything you are doing makes you a great instructor.

DO Identify Students That Are Ready to Learn

Make a conscious effort to identify the right students for each course you teach. Set yourself up for success by teaching the right people the right things. When you are preparing to provide training, ask enough questions to make sure you understand the audience and what will make them willing and ready to learn from you.

Develop with a Strategy

DO Demonstrate the Value of Training

Effective training is one of the most powerful tools a company has to increase revenue, avoid costs, or improve services. Make sure your training does that, and make sure you can demonstrate it.

DO Improve Your Training from Good to Great

Start with the list in Chapter 6 and build on it. You know your product, your customers, your industry, yourself, and your company. Be systematic in your approach to improve each training class. *Great* is limitless. *Effective* is not a destination that you reach to find rest. You can always improve, always get better, and always be a little more awesome.

DO Inspect and Evaluate Your Training

If you are always going to be improving, you must always be inspecting, or evaluating. Just as your product improves by getting feedback from your customers and others, so does your training improve through good feedback. Ask questions that you can act on and from which you can learn. Make your evaluations meaningful.

Design with a Structure

DO Dethrone King Content

Content is too important to put first. Make it even more powerful by letting it serve learning.

DO Use the 4×8 Proficiency Design Model

Use any model that helps you properly evaluate the business goals, audience, objectives, and learning activities before creating the content. Your content will become more powerful as a result.

DO Build Engaging Content and Deliverables

Use the right tools to deliver your content. PowerPoint slides are effective as a visual aid when facilitating a training class, but PowerPoint wasn't designed for instructors to deliver content by reading it like a book. Create your student guides and handouts to engage your students and keep them active in their learning process.

Deliver with a Purpose

DO Speak Up

Continuously improve your public speaking skills. Don't allow a fear of public speaking to keep you from facilitating, but don't let facilitating keep you from improving your public speaking skills either. Work on both decorative and declarative speaking, with the goal of being an even better instructor.

DO Shut Up and Listen to Your Students

Involving your students in their own learning requires listening to them. Ask them questions, let them answer questions, watch them learn, and watch them perform. Observe them for true learning. When they aren't absorbing it, slow down and say it again. Teaching requires a two-way conversation and two-way conversations require listening.

DO Stand Up and Be Confident

Demonstrate poise and control of your classroom. Have fun and look like you're having fun! You are in control and you are the expert. There is nothing to fear and your students will appreciate your confidence. Confidence is not egotistical. You can still be humble and confident, and you should be. Students will learn more from an instructor that confidently says "I don't know" than from the one who timidly makes up an answer.

DO Prepare for Difficult Students and Circumstances

You will have difficult students. Prepare for them. How you handle difficult students and circumstances can be the difference between effective and ineffective.

DO Deliver Effective Virtual Training

Bringing proficiency learning to a virtual setting is challenging but rewarding. Stretch yourself to try new techniques or tools, especially when the goal is training and not presenting.

DO Deliver Effective Technical Presentations

Delivering information and increasing one's knowledge about a topic—and doing it well—is still important. Don't assume that hands-on learning means you don't have to be an effective presenter. Just be certain whether a presentation is what is needed. Don't offer training when a presentation would be more effective and don't create a technical presentation when you want to change a skill.

DO Allow for Flexibility When Training in Other Cultures

There are company cultures, geographic cultures, and country cultures—all kinds of cultures in which you may have the opportunity to experience and teach. Embrace the opportunity, allowing for some flexibility as you do so. Demonstrate an understanding for people and a desire for them to internalize the training by making it easy for them to associate with in their context and culture.

Don't Forget the Details

DO Define Certification Properly

Prepare a list of requirements for certification. This will help you define it properly—and help others understand the difference between a certificate course and a certification course.

DO Manage the Details Properly

Behind every effective training program is an administration process that helps trainers and managers measure success, maintain consistency, and track students and instructors. The details are important to get right. It is necessary to know what must be tracked and define a plan that works for you and your company.

DO Mentor New Employees

Mentoring can be a key factor in the success of a product training program. Mentoring provides an opportunity for subject matter experts to make a difference in the life and career of a new or junior employee. When careful attention is paid to attitudes and outcomes, it can be a powerful program that benefits the mentor, the mentoree, and the company.

Conclusion

If you *do* all of these things, you will be well on your way to becoming proficient in delivering proficiency. You cannot be proficient in something you don't do. Practice, practice, practice. You will find, as I still do, that you will get better every time you teach. The goal of practice is to improve one's default abilities. Even the best instructors will fall back into their default comfort zone under less-than-perfect circumstances. Only effective practice can ensure that your default comfort zone is an effective training ability.

As you help to increase your students' value by improving what they can do with your product, you will have fun and you yourself will become a better employee in the process. The personal rewards for you are endless. You, too, will become more valuable to your company as you demonstrate an ability to navigate the process of transferring your knowledge to others.

You will find that little is as rewarding as seeing people change the way they do things because of your teaching. You may want to keep practicing to continue training. You may even want to become a professional trainer. Because, when you follow the principles in this book and practice them diligently, you will discover that you **DO deliver effective learning**.

Making It Practical

Your students invest time and money to learn from you. It is only worth it if your teaching changes what they DO. In the same way, you have invested in reading this book. It is only worth it if you go out and DO something differently. Do not try to do too much at once. Set a goal and prioritize your action items.

1) List at least one thing from each of the 20 items listed in this chapter and choose one of them to improve on next time you teach.

2) Prioritize the list above. Pick one that you will start working on today.

Part VI

For the Boss: Executive Overviews

22

The Foundation of Hands-On Learning

An Executive Summary

An Overview

Proficiency is the ability to consistently do something well. That is the goal of product proficiency training. It is doing, not knowing, that makes your employees valuable. Knowledge is only powerful as it increases our ability to do something with it. The same is true of students learning about your products. Increasing their knowledge will only help them if they are able to convert that knowledge into action.

In order to help others convert knowledge into action, trainers need to appreciate and understand how adults learn. Adults learn best when they can learn *while* doing. It is common, in instruction, to spend a significant amount of time learning by hearing and seeing before ever being able to touch the product. Whenever possible, instructors should let students get their hands on the equipment as early as possible. This change of approach is more than just an engagement technique. It improves learning. The goal is not for students to hear a lot about your product; the goal is for them to be able to do something differently with your product 6 weeks, 3 months, and a year after the training session.

This book further develops the way instructors turn that knowledge into action in Chapter 2. They do it by building on the experiences and backgrounds that all students bring to the classroom. Adults learn by constructing knowledge for themselves. Learners do that by taking previous knowledge and similar experiences and adding to them. Good instructors learn how to take the experiences of their students and help them build knowledge on that foundation.

Proficiency training requires experience-based activities. The true value of hands-on learning is learning that makes a difference. Lecture can be helpful, but instructors should minimize it. In its place, students need to move through the three phases of hands-on product learning (exhibit, execute, and explore) so that they can develop a proficiency with your product or solution. It is not enough to watch the instructor or other students perform a task, even if the task seems simple. They may understand the concept, but they will not be developing proficiency. They cannot be proficient in something they have not done. If a picture is worth a thousand words, experience is worth ten thousand pictures.

Product Training for the Technical Expert: The Art of Developing and Delivering Hands-On Learning,
First Edition. Daniel W. Bixby.
© 2018 John Wiley & Sons Ltd. Published 2018 by John Wiley & Sons Ltd.
Companion website: www.wiley.com/go/Bixby

A critical chapter for subject matter experts is Chapter 3. This chapter begins to reveal to the reader why their own abilities with your solutions make teaching difficult. It takes skill to teach skill. It takes a *different* skill. Just because an individual is skilled or educated in your product does not mean they know how to teach others about it. In fact, the more they know about your product, the more natural they are in using it or performing certain functions with it. However, the more natural they are in using your product, the more unnatural it will be for them to teach about it. Your best and brightest subject matter experts are like athletes who can perform tremendous feats and do not even know how they do it.

When technical experts begin to grasp why teaching is difficult for them, they start to enjoy the challenge. They realize that training really is another skillset that they can develop and add to their list of qualifications. Many experts have taught a training class in the past, but few experts have learned how to teach. With a little coaching and an understanding of the four levels of competency, technical experts can become effective instructors. They must learn, however, to rid themselves of assumptions and become more aware of what they really know. That is key—*they must be aware of what they know*. Most specialists have moved beyond awareness into a state somewhat like mental muscle memory, where they just do it because—well, because they just do it. But it is impossible to teach what you don't know that you know.

Chapter 4 deals with the students, the other part of the learning relationship. Technically speaking, no expert can transfer his or her knowledge to anyone. Learning is something that only individuals can do. This concept helps instructors understand why different students can process the same material so differently.

In order to learn, students must be ready to learn. There are four principles that govern their readiness to learn. First, students must recognize the need for learning. Second, they must take responsibility for their learning. Third, they must relate what they are learning to their own experience. Fourth, they must be ready to apply it.

As instructors learn how to determine where students are in regard to those principles and heighten the students' own awareness of them, they become instructors that are more effective. As they develop an awareness of how adults learn and how experts can teach, they develop an educational approach that is critical for the success of any training program. They understand that training is only effective when it changes a person's behavior or—in the case of product training—the solutions one provides using your products.

How You Can Help

As the manager, there are certain things you can do to help your subject matter experts be instructors that are more effective. The most important thing you can do is to emphasize effective learning. Ineffective training is expensive. Training that merely tells people about your product, but does not change what they can do with your product will cost you and your customers money and have very little return on investment. Let your product experts know that effective training is important to you, and then back it up with these important actions:

1) **Help them improve their skills**

   Do not ask your product experts to train unless they have learned how to be an instructor. Never assume that because someone has the knowledge or is proficient with your product, they will be an effective instructor. In fact, do not assume that because they carry themselves well in front of people, they will be a natural instructor. Some are more natural than others are, but make sure they know how to teach before giving them that responsibility. Investing in their ability to train effectively will save you money very quickly and be professionally rewarding for them as well.

   As with any skill, the most effective instructors are those who instruct regularly. It is better to find a few that do it well and allow them to sharpen those skills by teaching more often than to expect a larger number of people to become proficient in something they rarely do.

   When you do choose an expert to teach others, let them know that training is important to you. Send them to conferences and let them join professional societies or other groups that will help them continuously improve.

2) **Allow them to limit the class size**

   In order for proficiency training to be effective, instructors must keep class sizes small enough to allow all students to perform the tasks and exercises required to meet the objectives of the course. That will vary by product and objective. Listen to your instructors and allow them to make that determination. Give them the authority and backing to limit the number of students to an effective class size. The product quantity should be proportional to the class size.

3) **Give them the products required for hands-on training**

   Pictures and cardboard cutouts are not sufficient for hands-on training. One demo product that students gather around to watch the instructor do something with is not hands-on training. Help your instructors develop training that allows all the students to perform the required tasks with the equipment by making sure they have access to that equipment. Knowing that students leave the class with experience—being able to say "I did it," not just, "I saw someone do it"—will make the investment worth it.

Conclusion

Changing the way your company defines successful training needs to start at the top. Embrace and encourage a philosophy that combines action and knowledge. Put an end to the philosophy that delivers information with a hope that some of it will turn into action. No matter how excited your trainers are about teaching *while* doing, they are going to need your help to make that change a reality.

23

The Strategy of Hands-On Learning

An Executive Summary

Overview

Just as knowledge can be powerful when students put it into action, a great philosophy also becomes powerful when instructors have a strategy to execute it. This is the driving message in Part two.

Chapter 5 provides an overview of a business strategy and may be helpful for executives to read and absorb. There are at least three main ways to run training as a business. One way is to operate it as a profit center. Usually that involves selling training seats. The more seats you sell, regardless of the audience, the more money you make. For some companies, this approach can be successful. For most, it is not, since the "profit" they are making amounts to barely covering their costs. As a rule, if you would not run a product line for the profit you are making through training, you should not run it as a profit center.

Companies can also run training as a cost of doing business. Sales people generally prefer this option, since it helps to limit the total cost of ownership for their customers. You should always have a price associated with training, even if you choose to give it away to certain or all customers. This helps to maintain the value of the training program. It may also lead to a hybrid approach. In this third option, managers run training as a cost center while submitting the payments received for training as an offset to the budget.

Regardless of the business strategy, the true value of training is what should be calculated, measured, and reported. If you are a product manufacturer and your company exists to make money by selling a product, the training department exists to do the same. Effective product training will increase sales. Resellers sell more of what they know better. Systems designers integrate the products with which they are most familiar. Consultants refer products with which they have the most experience. This is the true value of training, not how many people took the class. When you provide effective training to your channel partners, you will find that training is one of the most powerful sales tools you have. Technical instructors have the awesome responsibility of getting your partners to be more proficient in using your product than your competitor's product. When that happens, your partners will lead with your product.

Product Training for the Technical Expert: The Art of Developing and Delivering Hands-On Learning,
First Edition. Daniel W. Bixby.
© 2018 John Wiley & Sons Ltd. Published 2018 by John Wiley & Sons Ltd.
Companion website: www.wiley.com/go/Bixby

Chapter 6 takes the instructor through nine comparisons that will help them turn their good training programs into great ones. Different programs may require a different focus from this list. It may vary, depending on the maturity of the program and the reasons for which it was developed. The nine focus areas are as follows:

1) Aim at the right target: Doing versus knowing
2) Change the approach: Facilitator versus lecturer
3) Call it the right thing: Training versus presentation
4) Make it sustainable: Standardized versus customized
5) Measure the right things: Performance versus reactions
6) Value the right things: Results versus head count
7) Use the right delivery method: Effectiveness versus availability
8) Continue the conversation: Process versus event
9) Keep improving: Progress versus contentment

Chapter 7 is an introduction to evaluations, both for the student and for the class. Exams can help to evaluate whether the student met the course objectives. Surveys can help evaluate whether the class is meeting the goal. They should be more than just a reaction to the class. Follow-up surveys help to determine if students are retaining and using the knowledge they gained in the class. Results from those surveys can significantly improve future classes.

How You Can Help

An instructor cannot build a strategy alone. It requires the authority of senior leadership to implement the strategy effectively. Many strategy items require acceptance and agreement from multiple departments and leaders. For that, they need your help. Little will frustrate a great trainer more than to know what the right strategy is, but not have the support to make that strategy happen. Here are three things you can do to help:

1) **Help them calculate the true value of training**
   Many trainers do not have access to the sales numbers required to make these calculations accurately. If they do, they may need some business guidance to make sense of the data. Help them get and interpret the information they need. Emphasize, across your company, that training is about revenue growth, cost reduction, or improving customer satisfaction, just like any other department in the company.
2) **Insist on using the right terminology**
   This can be a difficult task in some companies. Many events are often called "training" when, in fact, they are not (Table 23.1).
   If the training must be measurable, sustainable, and traceable, make sure the programs are run by a training department that can do all three.
3) **Let them choose the right delivery method**
   Good instructors know how to determine if their students can meet the objectives of the training course. They know if students will be able to meet the objectives in an eLearning environment or if meeting the objectives will require face-to-face instruction. Whatever the decision, do not ever sacrifice the outcome to simply meet a predetermined delivery method. eLearning (online learning) and mLearning (mobile learning) are wonderful tools that companies should use wisely. Using them ineffectively only dilutes their power. Good training programs will start with the objective and let that determine the delivery method, not the other way around.

Table 23.1 Presentation versus training chart.

Presentation	Training
• Successful when it changes or enhances knowledge—about knowing	• Successful when it changes or enhances a skill—about doing
• Motivates and encourages change	• Demonstrates change
• Audience size is irrelevant	• Learning is individualized
• Validation can be immediate	• Validation may be lengthy
• Focus is on the delivery; how it is *presented* and builds on the presenter's experience	• Focus is on the receiving; how the learning is internalized and builds on the student's experience
• Often *is* the learning event	• Generally it is one step in a process

Conclusion

Effective training can help you and your company be successful, but it starts at the executive level. Instructors need your leadership to help create an effective strategy of proficiency training. Make effective training an important part of your growth strategy. Help your trainers improve their skills to make it your most powerful sales tool. Most subject matter experts want to help others develop the same skills that they have. Help them help others.

24

The Structure of Hands-On Learning

An Executive Summary

Overview

Designing training correctly is critical for the success of any training event or program. Product experts must be involved in designing product training. Their involvement, however, is often limited to delivering content. Content is key, but it is not king. This section helps subject matter experts understand content's role in the design process.

Chapter 8 is a pivotal point in the book. Most experts, when asked to prepare for teaching a class, will immediately begin compiling the content. That is similar to finding out you are going on a trip and immediately packing the car. There are important things that the traveler must find out prior to packing the car. They must know where they are going, who is going with them, and what they will do when they get there. Only with those questions answered can one pack the car with confidence that they are bringing the correct material.

It is also important for experts to question whether training is the solution to the problem or not. If a salesperson sells your product as a solution to a problem it cannot fix, it will make your product look bad when it fails. The same is true with training. The worst possible advertisement for training is to promote training as the solution to a problem it cannot fix. There are many problems training cannot fix. Quality issues, motivation issues, pricing, and so on, might all be reasons a product is failing or not selling. Training may not be able to fix those.

Chapter 9 explains the design process. Introduced here is the 4 × 8 Proficiency Design Model (Figure 24.1). It consists of eight steps in four layers. Designers perform the layers sequentially while completing the steps within each layer in any order. This model was designed specifically for product training and is based on other good design models that ensure that the right content gets to the right people. The model is also based on a philosophy of learning that incorporates doing, not just listening, into the learning process.

Chapter 10 is a practical chapter that gives potential instructors points on making a student guide and visual aids like PowerPoint slides. Student guides are a helpful addition to a training class. A student guide is more than just the instructor's slides printed out for the students. The purpose of the guide is to enhance the student's learning

Product Training for the Technical Expert: The Art of Developing and Delivering Hands-On Learning, First Edition. Daniel W. Bixby.
© 2018 John Wiley & Sons Ltd. Published 2018 by John Wiley & Sons Ltd.
Companion website: www.wiley.com/go/Bixby

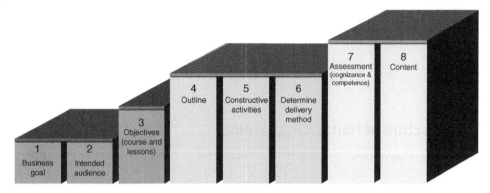

Figure 24.1 4×8 Proficiency Design Model.

experience, adding information the instructor may not have time to address, helping the students take the correct notes, and emphasizing the objectives of the course.

PowerPoint slides, or any equivalent visual aids, should help the instructor make a point. Instructors should not create slides that look like a technical manual. Graphics are better than words, and fewer words are better than complete sentences. Trainers should be familiar enough with the material that complete thoughts are unnecessary. If students need them, they should be included in the student guide.

Whenever a new curriculum is developed, it is important to test the class with a pilot class. Only after an instructor has the opportunity to teach a pilot class, will he or she be able to determine the necessary length of the class and if the exercises help the students meet the class objectives. Up until that point, all timelines are guesses and exercises are suggestions.

How You Can Help

Getting the structure right is important and as a decision-maker, it requires your consistent help. Mediocre training is inevitable if executives across your company do not understand and support a good design process. Here are a few specifics you can do to help:

1) **Make sure training is the solution before committing to it**
   Nothing will more quickly ruin a training program than good training that is ineffective. This can happen when curriculum designers create good training that is not the solution to the problem. It is tempting to suggest training as a solution quickly, especially if one misunderstands the investment required to design it correctly.

2) **Make learning more important than content**
   Content is most valuable when trainers can deliver the right content to the right people. Make sure that you allow your instructors the time necessary to get the content right. Developing the right curriculum takes a lot of time—about 40 hours of preparation per hour of training delivered. By supporting your trainers and giving them the time required to go through a valid curriculum design process, you will be supporting real learning over merely the delivery of content.

3) **Attend a pilot class**

By getting involved in the training process, you show your support for effective learning. Attending a pilot class will communicate that you support learning and understand what is required to perfect the learning process for your customers. Your attendance will also help you to see the instructors in action and encourage their efforts.

Most importantly, your presence signals a support for continuous improvement. A pilot class will not be perfect. In fact, the goal is to work out imperfections. Your presence during those potential imperfections will do more to support improvement than waiting until they have finalized the class.

Conclusion

Few subject matter experts understand the importance of designing their classes correctly. Of those that do, even fewer know how to do it. This section will help them understand both the impact of well-designed training and the process of doing so.

25

The Facilitation of Hands-On Learning

An Executive Summary

Overview

There is a subtle but distinct difference between presentation skills and facilitation skills. Many employees have taken classes designed to improve their presentation skills. Continuing to improve those skills will certainly increase their ability to teach others. More importantly, however, is the ability to facilitate training. Generally, facilitation requires a smaller, more intimate audience, and necessitates significantly more student engagement.

Chapter 11 is a written overview of how instructors can engage their students verbally. Two different skillsets are necessary concerning verbal communication: decorative language and declarative language. Decorative language deals with *how* we say things, while declarative language deals with *what* we say. It is important to control both. A chart and exercise is provided to help instructors identify which of the six decorative areas they need to work on (energy, breathing, pitch, tempo, volume, and articulation) and which of the four declarative areas they have strengths or weaknesses in (jargon, grammar, word crutches, or silence).

Chapter 12 helps instructors understand the importance of engagement. Engagement happens when the instructor refrains from speaking, more than when the instructor is lecturing. Students need time to absorb what the expert is teaching them. Learning happens at the rate of absorption, not at the rate of hearing or speaking. Instructors teach best by asking questions (lots of them!) and getting students to ask them as well. Except in rare cases, it is best to encourage questions throughout a training class, rather than adding a question and answer time at the end.

Nonverbal communication is the most important element of teaching, when it comes to whether our students will believe us or not. Chapter 13 helps instructors learn the importance of posture, facial expressions, eye contact, gestures, physical presence (sometimes referred to as movement and spatiality), and physical appearance. This chapter reminds readers that they are teaching individuals and they need to make each person feel like the primary audience. They can do this and have the added benefit of controlling the classroom, by teaching from different places and moving into the near presence of students as they teach.

Product Training for the Technical Expert: The Art of Developing and Delivering Hands-On Learning,
First Edition. Daniel W. Bixby.
© 2018 John Wiley & Sons Ltd. Published 2018 by John Wiley & Sons Ltd.
Companion website: www.wiley.com/go/Bixby

This practical chapter also addresses the difference between the observed communication factors mentioned previously and other perceived communication elements. Instructors should be genuine, likeable, available, prepared, confident, and in control. This control starts the instant a student walks through the door. If the instructor is still preparing, or struggling with equipment, she is not available for her students.

There are difficult students in almost every class. Good instructors are prepared to deal with them. Chapter 14 helps prepare instructors for some of the more common difficulties they may face.

Chapter 15 helps the reader apply the previous principles in a virtual setting. Instructors will likely need to teach others via a live webinar or phone call. When they can deliver the objective in a virtual setting, it can be a great cost-saving tool. Engagement is still important, and this chapter provides practical tips for engaging a virtual audience.

Presenting knowledge is still an important task that must be taken seriously. Chapter 16 provides a five-step process to help experts develop technical presentations that are effective. Just as importantly, it helps them determine when to use a presentation to deliver technical knowledge.

Many instructors have the opportunity to travel internationally to teach about your products. When they do, it is helpful to be aware of cultural differences. Sometimes cultural differences are exaggerated, but often they are real. Instructors need to know what may change across cultures and what does not. The key is flexibility. Training through an interpreter, or even just training those learning in a second language, will take more time.

How You Can Help

Facilitating skills are as important to a trainer as closing skills are to a salesperson. Understanding and advocating the value is one of the best things you can do as a leader. Here are three specific things you can do to help your product experts facilitate learning more effectively:

1) **Give them the experience and leadership necessary to improve**
   Facilitating training is a skill. The only way to get better at a skill is to practice that skill. However, practice does not make perfect, in spite of what you may have heard in the past. Practice makes permanent. If your instructors practice a lot, but do not have an opportunity to get feedback and help to improve, they may make bad habits permanent. Allow those who are naturally inclined towards training to get the necessary assistance to continue improving.
2) **Acknowledge the effort and time required for effective facilitation**
   Training should never be anyone's secondary job, at least in terms of importance. Even if training is not what they normally do, it is still a primary task. Training is important, even essential for the continued growth of your product. If you have an experienced expert who is willing to share his or her knowledge with others, you need to acknowledge them.
   When instructors must travel to teach a class, allow them the time necessary to be effective. They need to arrive early and be in their best possible frame of mind and body in order to make the training as effective as possible.

Never ask an expert to teach a training class without allowing them to clear their schedule. Experts cannot teach effectively as an addition to their other daily activities. Know that for every hour of effective training delivered, they likely spent many more in preparation.

Your gratitude and awareness of the impressive amount of time it takes to deliver effective training is an important part of the process.

3) **Visit their classes**

Trainers are hosts. They have a lot to think about, and a lot of things are happening at the same time. They sense a huge responsibility to make their students happy. They want their students to enjoy their time learning.

You can help. Surprise them with a little extra treat or just stop in to tell the class how much you appreciate them being there and how much you appreciate their instructor. Mention something that she or he has done. Make them proud. It will go a long way toward giving them the extra energy they need to make the class successful.

Conclusion

If you already provide effective training and can prove it, congratulations. If you do not, know that you can. The principles in this book offer a starting point and discussion topics. Reading the book will not create a great instructor. Someone said that you cannot learn to swim in the library. The same is true for training. Start putting these principles into practice, and little by little you will develop a training program that works. You will find, as countless others already have, that technical experts can deliver effective learning. That learning will drive real and lasting business results.

26

The Operation of Hands-On Learning

An Executive Summary

Overview

There are many operational tasks required to run a product training program that is both efficient and effective. This section covers three areas that may or may not be part of the direct responsibility of the subject matter expert. Even if an expert is not account-able for the outcome of these areas, it is important that they understand how each one functions and how it affects their training, their program, and your company.

Chapter 18 covers an area that many delve into with little or no knowledge: product proficiency certification. Certification is an important topic in product training. It is important for instructors, executives, and many in between to understand the difference between a certificate and a certification. There are good reasons to offer a certification program, and companies should not be afraid to do so. Product profi-ciency certification is an official acknowledgment of an individual's ability to perform a particular task with your product. General industry knowledge is important and may be a prerequisite to certification, but certification requires evidence of a particular ability. A certificate, on the other hand, signifies the successful completion of an education program or process.

Chapter 19 is an overview of how to administrate a product training program. The message to the product expert is that administration matters. Never take lightly the value of good administration.

The chapter starts by covering three areas that are important to manage in any program. Administrators must make sure that every training program is measurable, sustainable, and traceable. Executives need to be aware of the dangers of providing train-ing without tracking the recipients of that training. Too many experts provide "training" on your product, and some may even give out pseudo-certificates with your company logo on them without your knowledge. Companies can and have been held liable for training delivered, so it is important to ensure that all training events are tracked properly.

This chapter also covers how to track curricula for changes and make sure that all instructors are teaching the same course. Consistency of teaching methods and outcomes will provide more accurate metrics to determine the value of a course.

While it is not the main emphasis of the chapter, it does provide administrators with some ideas for measuring the business value of training as well. Note that I encourage

Product Training for the Technical Expert: The Art of Developing and Delivering Hands-On Learning,
First Edition. Daniel W. Bixby.
© 2018 John Wiley & Sons Ltd. Published 2018 by John Wiley & Sons Ltd.
Companion website: www.wiley.com/go/Bixby

that as a six-month project every few years, since it takes a significant amount of time and resources to get accurate information.

Another area of indirect concern for many product experts is what is often referred to as "transfer of knowledge" from senior employees to more junior employees. The concern is very real for talent development experts and a top priority for human resource departments. One of the best ways to make that happen is in an informal mentoring relationship.

Mentoring won't happen unless the experts believe in it—in its benefits to them, to the new employee, and even to the company. Mentoring fits well with a hands-on approach to learning, because it is grounded in learning *while* doing. This chapter encourages participation in stretch tasks while assisting a more experienced expert. Learning that matters and that happens in a real environment with real and important results is more effective than any other type of learning.

Just because mentoring encourages informal learning does not mean the program itself should not be organized or formalized. Every company will have different ways to do that, but some ideas are provided in this chapter.

Every good training, presentation, or book has a call to action. This book is no exception. The last chapter, *Now, Go Do It,* is the instructor's call to put what they have read into action. They cannot become a good trainer until they start training. The sooner they do it, the better. This chapter may also serve as a summary of the book, and one they can review from time to time.

How You Can Help

First, thank you for reading these executive summaries. Your support for effective training will pay dividends. Here are a few things you can do to help make any product training program more effective:

1) **When certification is required, make it meaningful**

   Make certain that certification programs add value to your customers. Make them more than just an agreement between companies. Product proficiency certifications are for individuals who have demonstrated an ability to do something with your product. Hold to the standards of certification, for legal and practical reasons, and help your instructors do the same.

2) **Get good administration help**

   Don't expect instructors to be experts in the administration of training, experts in the design of training, experts in the delivery of training, and experts in your product. Too many great trainers get burned out because they are asked to do too much. Administration is an important task. Asking trainers to act as administrators or adding it on as a subtask to another office manager's position is only sending the message that it doesn't really matter. Make sure that those in charge understand the important role they have and how they can add value to the training program.

   Invest in the administrators as well as the trainers. Administrators can benefit from conferences and learning opportunities. Give them the formal training in the learning management system or other tools that they need to be successful.

3) **Develop a mentoring program**

Mentoring will happen. Junior employees will watch and learn from veteran employees. Help them learn what they are supposed to learn and what your organization wants them to learn.

If there is a culture of hoarding or hiding information, start the process of eliminating that cancer. Emphasize teamwork and the human value of helping others. Reward a behavior that exemplifies collaboration. Become a mentor yourself and don't be afraid to state what you are learning from the mentoree. This will set a good example to follow.

Conclusion

Effective training is artwork as a team sport. Some instructors will be better facilitators than others. Some designers will be able to work with subject matter experts in a way that makes designing come together more quickly than others will. To deliver training that is really going to make a difference, your company needs product experts, instructional experts, curriculum design experts, administration experts, talent development experts, and more.

You can teach a salesperson the basics of selling or a methodology that works. But much of their success will come in the art of selling. The same is true of product training. It is not a math process that has only one correct answer. Training is an art form, and some will be more artistic than others. Find trainers that love people and love teaching. If they want to make training their primary job function, encourage it, if at all possible. The more they do it, the better they will get at it. When you have instructors, designers, and administrators that encourage hands-on learning, you can have a product training that is effective and changes what your customers or employees can do with your product.

Index

Product Training for the Technical Expert: The Art of Developing and Delivering Hands-On Learning,
First Edition. Daniel W. Bixby.
© 2018 John Wiley & Sons Ltd. Published 2018 by John Wiley & Sons Ltd.
Companion website: www.wiley.com/go/Bixby

Printed and bound by CPI Group (UK) Ltd, Croydon, CR0 4YY

27/10/2024

14580362-0002